GEAR HOBBING, SHAPING AND SHAVING
A Guide to Cycle Time Estimating and Process Planning

By
Robert Endoy

Richard Perich
Publications Administrator

Published by
Society of Manufacturing Engineers
Publications Development Department
One SME Drive
P.O. Box 930
Dearborn, Michigan 48121

GEAR HOBBING, SHAPING AND SHAVING
A Guide to Cycle Time Estimating and Process Planning

Copyright © 1990
Society of Manufacturing Engineers
Dearborn, Michigan 48121

First Edition

First Printing

All rights reserved including those of translation. This book, or parts thereof, may not be reproduced in any form or by any means including photocopying, recording or microfilming or by any information storage and retrieval system without permission of the copyright owners. The Society does not, by publication of data in this book ensure to anyone the use of such data against liability of any kind, including infringement of any patent. Publication of any data in this book does not constitute a recommendation of any patent or proprietary right that may be involved or provide an endorsement of products or services discussed in this book.

Library of Congress Number: 90-61098

International Standard Book Number: 0-87263-383-7

Manufactured in the United States of America

FOREWORD

The theory of gears is based almost entirely on mathematical science. Lengthy and complex calculations are required to specify gear data, gear cutting tools, and gear inspection instruments. Gear manufacturing companies employ specialists in gear and tool design, who, with the aid of computer programs, generate data for gears and cutting tools. Many more manufacturing and industrial engineers are, however, involved daily in estimating and planning production facilities and rates for gear cutting operations. These activities require cycle time calculations, and extensive knowledge of processes and cutting parameters.

This book covers one aspect of gear manufacturing often neglected or only summarily mentioned in other works: how to estimate effective cycle times for hobbing, shaping, and shaving operations. Reliable estimating requires a substantial amount of calculations and a good understanding of processes, machines, cutting tools, and cutting parameters. A thorough discussion of all the variables that affect cycle time provide a better insight in gear manufacturing processes, their field of application, and their limitations.

Gear process planning is also considered a special skill. In addition to having a broad knowledge of gear manufacturing, the process planner must also be experienced in blank preparation techniques, heat treatment operations and their affect on quality, and finishing operations after heat treatment.

The wide variety of gear types found in drivelines implies that each gear is different and requires a multitude of manufacturing processes. Actually, gear process planning is based on a logical sequence of operations and principles, applicable to the majority of gears in automotive, truck, agricultural, and construction equipment.

The second part of this book reviews the basic steps in the planning sequence. It details the transition from rough material to finished, machined gear.

The theory is illustrated with practical examples taken from existing production applications. It includes a discussion of process plans for typical gear configurations as well. It is not the intention of this book, however, to repeat basic theory found in great detail elsewhere. For the reader new to gears, it would be beneficial to consult works on general theory.

TABLE OF CONTENTS

Part I: Overview 1

Chapter 1
Introduction to Hobbing, Shaping and Shaving 3

Chapter 2
Hobbing 27

Chapter 3
Shaping 49

Chapter 4
Shaving 63

Part II: Process Planning 75

Chapter 5
Gear Manufacturing Methods 77

Chapter 6
Spline Manufacturing Processes 89

Chapter 7
Special Gear and Spline Processes 101

Chapter 8
Blanking 105

Chapter 9
Heat Treatment 117

Chapter 10
Grinding 127

Chapter 11
Process Planning Examples 139

Chapter 12
Gear Terminology 173

Part III: Bibliography 177

Index 179

Part I

Overview

Estimating production rates of machining operations, on new or existing equipment, is an integral part of advanced manufacturing planning and cost estimating. The standard time of a machining operation consists of the cycle time which is increased with allowances for activities inherent to the operation such as loading and unloading the part, tool change, personal time, in process gaging, machine cleaning, etc.

The standard time is used to estimate machine capacity allocations and manning of the line. Accurate, reliable, and consistent estimating practice is therefore a valuable tool in planning new facilities, investment budgets, direct and indirect labor headcounts, and piece cost. Because of the importance attached to piece cost and investment forecasts, estimating of production rates is a crucial activity which deserves the utmost attention.

Estimating machining times requires, in essence, a good understanding of the working principle of the machine. It also requires insight of the elements that control the output of an operation such as feeds and speeds, length of cut, number of cuts. Gear cutting operations are more complex to estimate because tool travel paths need to be calculated. Also, the output of a gear cutting operation can vary considerably depending on the various capabilities of gear machinery, and by the type of tooling used for the operation.

To facilitate the task, vendors of gear cutting equipment have developed sliding rules and graphs to assist manufacturing engineers in estimating cycle times for hobbing, shaping, and shaving. Although accurate, sliding rules and graphs are tedious to use and not suitable for integration in CAD/CAM applications.

Rather than being a disadvantage, the mathematical aspect of gear estimating can be turned into an advantage when used in conjunction with computer software. Once the formulas to calculate cycle times have been translated into computer software the manufacturing engineer can concentrate his or her efforts in selecting the appropriate machining method and cutting parameters, and leave the calculations to the computer. With the proper theoretical background, software is readily developed and tailored to individual needs.

Cutting parameters can be found by consulting tables of recommended feeds and speeds for various materials. Most of these tables have, however, the disadvantage of being too general and giving a wide spread between maximum and minimum values. They also do not take into account variables, such as rigidity of part and fixture, which have a considerable affect on feeds and speeds. Since most manufacturers produce families of like parts, databases which are built on own experience usually give more satisfactory results.

1
Introduction to Hobbing, Shaping and Shaving

HOBBING

The Generating Process

Gears for power transmission applications are almost exclusively made with involute profiles. The involute curve is generated by the end of a taut line as it is unwound from the circumference of a circle. The circle from which the line is unwound is known as the "base circle."

Point P on line L (see Figure 1-1) generates the involute curve when line L rolls over the base circle. Line T through point P is tangent to the involute curve and perpendicular to L. The principle of hob generating is based on producing an involute profile by many successive cuts tangential to the involute profile, with a straight-edged cutting tool. The profile of the cutting tool, called a "hob," is shown in Figure 1-2.

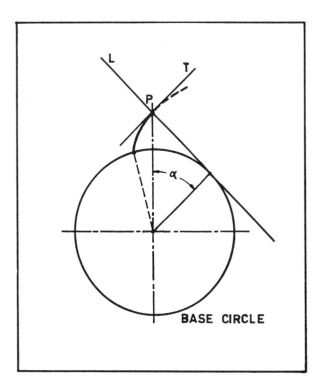

Figure 1-1. Involute curve generated by a point on a line which rolls over a base circle.

It is referred to as the "basic rack profile." Figure 1-3 shows how a gear tooth is generated on a hobbing machine by consecutive cuts of the hob. The advantage of involute gears is that tooth profiles can be generated accurately with inexpensive and universal tools. The proper function of an involute gear set is not affected by small variations in center distance.

Hobbing is a continuous gear generating method which can be best compared to the action of worm and wormwheel. The gear to be cut is represented by the wormwheel. It rotates in tight mesh with the hob, represented by the worm. The hob, which is a worm provided with cutting edges, is fed axially across the face of the workgear. The rotating speed of the hob is the hobbing speed. The rate at which the hob moves axially across the face of the gear is the hobbing feedrate.

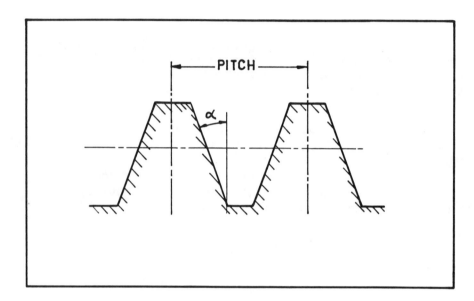

Figure 1-2. Basic rack profile.

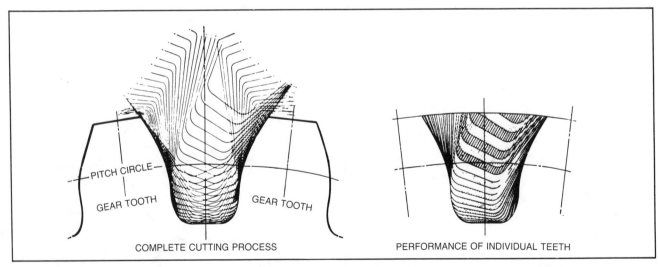

Figure 1-3. Hob generating action. (1)

(1) Drozda, Thomas J., Wick, Charles. Tool and Manufacturing Engineers Handbook, Fourth Edition, Volume 1: Machining. Dearborn, Michigan: Society of Manufacturing Engineers, 1983, p. 13-38.

Hobbing is the most common method used in the industry for production of spur and helical gears. It is the only available method of producing wormgears. Hobbing also can cut splines, sprockets, and most other forms that are uniformly spaced around an axis. Figure 1-4 shows forms, other than gears and splines, which can be produced by hobbing.

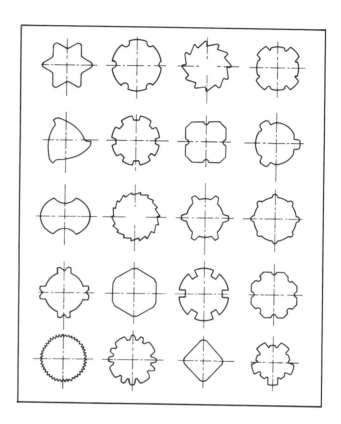

Figure 1-4. Forms which can be produced by hobbing. (2)

These forms are typically cut with single position semi-generating hobs, permitting the last engaged tooth in the hob to cut the radial face on the work. The projections or indentations need not necessarily be symmetrical about an axis that intersects the center of the work. The form also does not need to be of any special curvature. The only limiting factor is that the indentations or projections must be of sufficient width in proportion to height to roll the hob in and out without interference.

A single-position hob to cut an asymmetric form, such as ratchet teeth, must be accurately centralized relative to the blank. It cannot be shifted to a new position when the teeth begin to dull. A new centralizing position must be determined after each resharpening. High accuracy with respect to tooth spacing and runout are inherent to hobbing. Lead and involute profile accuracy is comparable to other generating methods. Surface finish of a hobbed profile is, however, not as good as the finish obtained with other methods such as shaping or shaving. Hobbing leaves a series of slight,

(2) Drozda, Thomas J., Wick, Charles. Tool and Manufacturing Engineers Handbook, Fourth Edition, Volume 1: Machining. Dearborn, Michigan: Society of Manufacturing Engineers, 1983, p. 13-35.

radial tooth marks. The width of the marks depends on the feedrate at which
the hob is fed across the workpiece (Figure 1-5). In some applications, it
may be necessary to remove these marks by a separate finishing operation.
Shaving or grinding are two such operations.

Hobbing Machines

Hobbing is a versatile method of gear manufacturing. It is applicable when
large numbers of identical gears are to be manufactured as well as for
single gears or small lots.

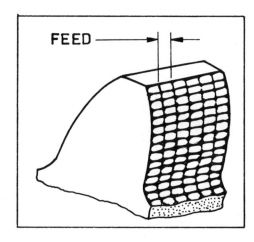

Figure 1-5. Typical hob feed marks on tooth flanks.

Change-over on a hobbing machine consists of fixture changes to locate and
clamp the component, and installing the appropriate hob on the hobbing
arbor. It should be noted that the same hob can cut spur as well as helical
gears with the same diametral pitch and pressure angle. The mechanism that
provides the incremental index to generate helical gears is called the
"differential" on a conventional hobber. The helix angle is obtained by
installing sets of differential change gears. The rotation speed of the
work piece, that produces the desired tooth spacing, is set by means of
change gears.

Computer numerical control (CNC) machines have eliminated the gear train
between hob and workpiece and replaced it by individual servo-drives for the
hob spindle and the worktable. As a result, machine settings are greatly
simplified on CNC hobbers and change-over time is drastically reduced.

Hobbing machines are produced in various sizes for parts within a specific
dimensional range. A 10 in. (254 mm) capacity machine is typically capable
of cutting gears to 10 in. (254 mm) pitch diameter, with maximum diametral
pitch 3, and 10 in. (254 mm) maximum face width. Larger machines can go up
to 118 in. (3 m) maximum pitch diameter, 0.85 in. (21.6 mm) maximum
diametral pitch, and 36 in. (914.4 mm) maximum face width. Most of the
hobbing machines used in manufacturing are vertical hobbers. The parts are
machined with the axis of the blank in a vertical position. Horizontal
hobbing machines hold the parts between horizontal centers. These machines
are used to cut tooth forms on long shafts. Horizontal machines often are
equipped with hollow spindles to accommodate very long parts. One typical
application of horizontal hobbing machines is to cut splines at the end of
long shafts.

Gear hobbers are extensively used in high-volume production of automotive gears. Uninterrupted production and high production rates are guaranteed by automated part load and unload systems.

The main advantage of hobbing versus shaping rests with the longer axial stroke of the hobbing head. Several parts may thereby be cut in the same cycle. Hobbing productivity can be tremendously increased by stacking two, three, or more parts on the fixture and cutting them in a single pass of the hob. By contrast, shaping is limited by the stroke length. Stacking of parts does not lead to more production output in shaping.

Hob Centering

Hob setting gages centralize the position of a hob tooth with respect to the axis of the blank. Normally, when cutting gears with a large number of teeth, there is no need for such accuracy. Centralizing with a hob centering gage is desirable when:

* cutting gears with a low number of teeth;
* using a hob having minimal number of flutes;
* angular relationship with another feature must be maintained;
* after resharpening of the hob;
* using semi-topping and protuberance hobs. Out of center conditions will result in asymmetric tooth profiles.

The hob always must be positioned so that the teeth doing the cutting are not so close to either end of the hob that the generating action would be incomplete.

Hob Shifting

Most gear hobbers are equipped with a hob shift mechanism that moves the hob over a preset distance along its axis after every cutting cycle. This shifting distributes the cutting action, and cutter wear, over as many teeth as possible before resharpening is required. It is used when cutting spur gears, helical gears, or infed worm gears. It is particularly useful in high-volume production when tool economy is essential.

In Figure 1-6, L1 and L2 represent the two extreme positions in which the

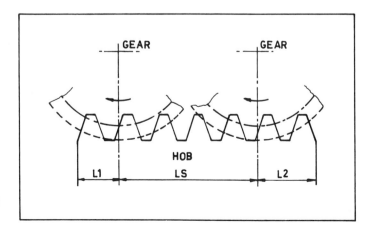

Figure 1-6. Available shift length (LS) between two extreme hob positions.

hob can be used for full generating action. LS is the total hob length available for shifting. A new hob is positioned in one extreme point (for instance L2) and shifted in the direction of the other point (L1). The available shift length LS is divided in a number of equal increments. The hob is shifted one increment after every cycle until the full shift length has been used.

The amount of hob shift is determined experimentally to give the longest tool life while producing good quality gears. For gears with a small number of teeth, it is necessary to maintain the hob centered. Thus, the shift amount is equal to the displacement needed to bring the next tooth in the center.

The hob can be shifted in two directions relative to the component. To understand the difference between methods A and B in Figure 1-7, it must be

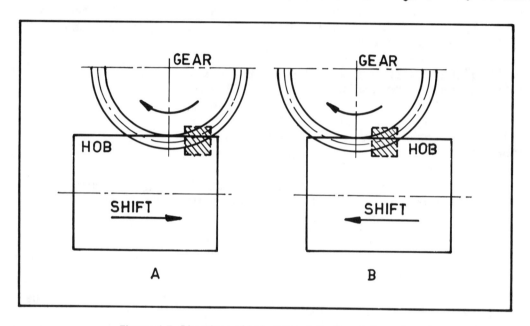

Figure 1-7. Directions of hob shift relative to rotation of gear.

pointed out that most of the cutting action occurs in a small area of intersection between hob and gear. This area is indicated by the cross hatch.

The hob is shifted in the opposite direction to the rotation of the component (A). The new sharp hob teeth that move into cutting position finish the gear teeth and produce a good machined surface. The dull teeth of the hob move to a position where most of the rough cutting is done, and the hob will wear more rapidly.

The hob is shifted in the same direction as the rotation of the work gear (B). The dull teeth move in finishing position, while new sharp teeth move into roughing position. The result is lower surface quality but less hob wear. For gears which are finished by shaving, this method is preferred because of superior tool economy.

Climb and Conventional Hobbing

The terms "climb" and "conventional hobbing" refer to the direction of hob feed into the workpiece with reference to the table or the spindle nose. For maximum stability, it is recommended that the cutting forces be applied against the worktable or nose of the machine spindle. In conventional hobbing (Figure 1-8), the hob is fed into the work, moving toward the table or spindle nose, parallel to the blank axis. In climb hobbing (Figure 1-9), the hob is fed into the work, moving away from the table or spindle nose, parallel to the blank axis.

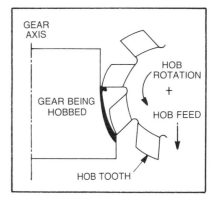

Figure 1-8. Conventional hobbing. (2)

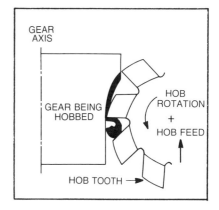

Figure 1-9. Climb hobbing. (2)

Both methods of feeding are equally applicable for either vertical or horizontal hobbing machines. Climb hobbing is advantageous for both soft and heat treated steels, particularly the latter. Greater feeds and considerably faster cutting speeds are possible. Climb cutting sometimes permits finishing soft steel gears in one cut, where two cuts would have been necessary with conventional hobbing.

In conventional hobbing, the chip starts thin and becomes progressively thicker. A squeezing or sliding action tends to occur as the hob tooth performs the cut. The end result is normally deteriorating cutting conditions and increased tool wear. However, this tendency to smooth can be used to advantage in finish hobbing of gears with low stock removal, (e.g. two cut cycles). Conversely, in climb hobbing, the chip starts thick and

(2) Drozda, Thomas J., Wick, Charles. Tool and Manufacturing Engineers Handbook, Fourth Edition, Volume 1: Machining. Dearborn, Michigan: Society of Manufacturing Engineers, 1983, p. 13-35.

becomes progressively thinner. This method avoids the sliding action and, as a result, is used mainly in the roughing of gears where high stock removal is required with good tool life.

Tests have shown that tool wear is more noticeable with conventional hobbing of coarse pitch gears than with the climb hobbing method. However, in the case of gears with high helix angles (28 plus), conventional hobbing gives better tool life. The better tool life is caused by the more favorable entry conditions between tool and workpiece.

Climb hobbing requires rigid hobbing machines maintained in good condition. The leadscrew and nut backlash are especially critical. Attempts should be made to eliminate backlash. Many modern gear hobbing machines use backlash-free recirculating ballscrews. The use of these ballscrews eliminates the effect of the axial feedscrew as a consideration for climb or conventional hobbing. Other factors, including gear material and rigidity of fixture and blank, influence the success of climb hobbing.

Hobs

A hob resembles a worm in appearance, with its cutting teeth on the outside of a cylindrical body following a helical path corresponding to the thread of either a left or right-hand worm. It is not strictly correct to say that an axial section of a hob is a rack. It is useful to think of a hob as a cylinder with a series of racks fastened on its periphery. Each rack is parallel to the axis of the cylinder (or nearly so). Each one is slightly displaced axially with respect to the preceding rack. As the hob rotates in timed relationship with the blank, each row of teeth successively cuts the next portion of the gear tooth spaces.

The selection of the proper type of hob for the cutting operation is based on analysis of accuracy and production requirements. Standard hobs are made to manufacturer's specifications and are available from stock. Standard hobs are used mostly for small lot production and prototype work. Hobs for high-volume gear production are usually made to special order. These hobs incorporate design features required to reach optimum performance and quality.

The basic dimensions of a hob are (Figure 1-10):

* Diametral pitch,
* Pressure angle,
* Spiral angle,
* Outside diameter,
* Length, and
* Bore diameter.

Diametral pitch and pressure angle are the two most important characteristics of the hob. They correspond to the diametral pitch and pressure angle of the gear to be cut. A hob of a given diametral pitch and pressure angle can cut a range of gears with same diametral pitch and pressure angle, but with different number of teeth. Spiral angle, outside diameter, length, and bore diameter are determined by hob design considerations.

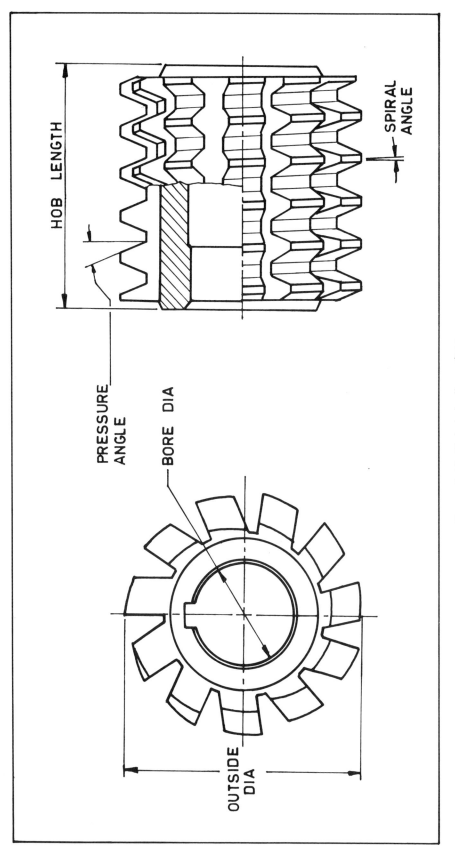

Figure 1-10. Main hob dimensions.

In addition to the basic dimensions, it may be necessary to specify one of the following features, or a combination of them.

- * Number of starts
    - Single start hobs
    - Multiple start hobs
- * Hob construction
    - Solid hobs
    - Inserted blade hobs
    - Indexable inserts hobs
- * Quality class
    - Class AA high precision ground
    - Class A precision ground
    - Class B commercial ground
    - Class C certified unground
    - Class D certified unground
- * Toothform
    - Basic rack
    - Protuberance
    - Semi topping
    - Root fillet radius
    - Tip relief
    - Root relief
- * Hob material
    - High-speed steel
    - Coated high-speed steel
    - Carbide
- * Work application
    - Roughing hobs
    - Finishing hobs
    - Pre shaving hobs
    - Pre grinding hobs

Number of Starts

Multiple start hobs are used to increase the productivity of gear hobbers. Significantly higher production rates can be obtained by using hobs with two, three, or more threads. However, gear design considerations and quality specifications restrict the application of multiple start hobs. Each case must be analyzed individually. Careful consideration must be given to the specified gear tolerances, quality and design of the hob, quality capability of the hobbing machine, and the maintenance condition of the machine.

Multistart hobs reduce the cycle time, and increase the output, of a hobbing operation because they allow the work piece to be rotated faster than with a single start hob. With a single-start hob, the work gear advances one tooth for each revolution of the hob. With a multistart hob, the indexing speed is increased proportionally with the number of starts. A two-start hob will double the rotating speed of the work gear. Since the machine cycle is inversely related to the rotation speed of the work, the cutting time will diminish proportionally to the number of starts.

The basic rule for application of multistart hobs can be expressed in two parts. First, the number of gear teeth must not be divisible by the number of starts in the hob. Second, the number of gashes in the hob must not be divisible by the number of starts.

When a gear with an even number of teeth (for instance 46) is cut with a two start hob, one thread of the hob will cut tooth spaces 1,3,5,7,...,43,45 and the other thread will cut tooth spaces 2,4,6,8,...,44,46.

Since the threads in the hob are ground in separate operations, each thread has its own specific manufacturing errors. Such errors will reproduce in the teeth of the component. The errors of thread one will reproduce in the uneven teeth. The errors of thread two will reproduce in the even teeth. It is clear that this condition downgrades the quality of the gear. To avoid this, the number of teeth of the gear and the number of starts in the hob must not be divisible.

If a gear with an uneven number of teeth (for instance 47) is cut with a two-start hob, then both threads of the hob will cut alternatively through the even and uneven toothspaces. For instance, in the first rotation of the gear, thread one will cut through the uneven toothspaces. In the second rotation it will cut through the even toothspaces. In the third rotation it will cut again through the uneven tooth spaces, and so on. The result is a gear with good tooth spacing tolerances but with a less accurate profile. The involute profile can however be improved in the shaving operation so that the use of a two-start hob does not cause an appreciable downgrading in quality.

Hob design must be such that the number of gashes is not divisible by the number of starts. Then, after every revolution of the work piece, the tooth is cut by a different thread and a different gash. The cutting marks will be displaced relative to each other resulting in a finer finish.

As seen, a two-start hob cuts two gear teeth in the same time that a single-start hob cuts only one tooth of the same gear. If both hobs have the same diameter and the same number of gashes, then only half the number of teeth of the two-start hob participate in the generation of one gear profile. The gear tooth is therefore generated by fewer cuts. The involute profile is less accurate, and the chip load per hob tooth is twice as big. A number of factors partially offset the productivity gains of multistart hobbing.

* To maintain good profile accuracy, multistart hobs are designed with more gashes so that the profile is cut by more teeth. To accommodate more gashes, the hob diameter is increased. The usable tooth length of the hob becomes shorter, with fewer regrinds per hob, and reduced tool life.
* Because of the larger outside diameter, the hob rotates slower, and the approach travel is longer.
* The higher chip load per tooth makes it necessary to reduce the feedrate to maintain acceptable tool wear.

Therefore, the output of the machine, although still substantially increased with respect to single-start hobs, is not a direct multiple of the number of threads. Generally, hobbing productivity increases 60% with two starts; 90% with three starts; 120% with four starts. Six threads is the physical limit for the number of hob starts.

Construction

Solid hobs are made from a solid chunk of high-speed steel. The teeth are relief turned, heat treated, and ground with a small diameter grinding wheel or with a pencil type grinding wheel. Because of the small grinding wheel diameter, grinding takes a long time resulting in higher cost. Solid hobs can be manufactured within very accurate tolerances (Class AA). That fact makes them especially suitable for precision gear cutting.

Inserted blade hobs are widely used in the gear industry because of their excellent tool economy. They consist of superior grade high-speed steel blades inserted in a body made of ordinary tool steel. The cutting blades are ground before they are installed in the body. They are mounted in a special grinding fixture and ground on a worm grinding machine using a large diameter grinding wheel.

The advantages of inserted blade hobs include the following points:

* Lower material cost since only the blades are made of expensive high-speed steel;
* Shorter grinding time resulting in lower manufacturing cost;
* More usable portion of the tooth length compared to solid hobs, resulting in more regrinds per hob and lower tool cost;
* The material structure and heat treatment of the blades can be controlled more precisely than with solid hobs;
* The blade material is more durable and wear resistant allowing higher feeds and speed and more severe cutting conditions.

Inserted blade hobs can be delivered in precision ground quality Class A for finish hobbing applications.

Quality Class

Hobs of quality class AA and A are used for finish hobbing operations when gears are cut to final part print specifications on the gear hobber without any subsequent finishing operations. Finish hobbing is mainly applied for large-diameter gears. In this case it is more advantageous to change the cutting tool between roughing and finishing cuts rather than moving the part to another machine for finishing.

Commercial-ground Class B hobs are used for pre-shave hobbing operations of gears which are finished by the shaving method. Ground gears are rough cut on the hobbing machine with Class C hobs. They are finished to close tolerances by gear grinding operations. Class C and D hobs are also used to produce other tooth forms--such as sprockets and splines.

Profile

A basic rack hob produces a true involute profile without any modification. This type of hob profile is not always used for gear cutting. Modified profiles are generally desired for optimum gear meshing. (Figure 1-11).

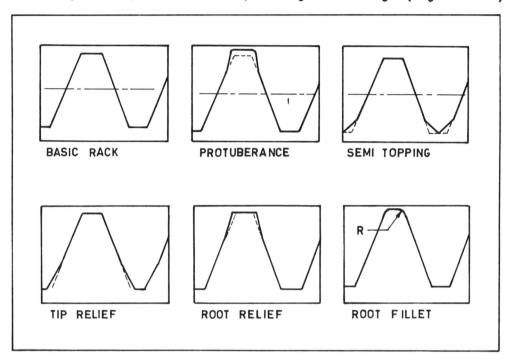

Figure 1-11. Various modifications to the hob profile.

Protuberance hobs cut a clearance in the root of the gear tooth, leaving finishing stock on the tooth flanks only. The purpose of the clearance is to facilitate the cutting action in subsequent finishing of the tooth flanks by shaving or grinding. It must be noted that protuberance hobs are not a necessary requirement for pre-shave hobbing. Shaving works just as well with non-protuberance hobs. In addition to being more expensive, protuberance hobs reduce the active profile length. They also reduce the contact ratio between gear and shaving cutter. Semi-topping hobs cut a chamfer at the tip of the tooth profile. This chamfer, called the "tooth tip chamfer," is often incorporated in gear design as it protects the gear teeth from handling damage. Accidental contact damage between gears (such as nicks) occurs in the chamfer area rather than on the tooth flanks. Since the tip chamfer reduces the active length of the involute profile, it should be kept as small as possible.

Tip relief hobs remove more metal near the tip of the gear tooth to permit interference-free tooth engagement between mating gears. Tip relief is used in heavy loaded gears to compensate for tooth interference. It is usually a cure for gear noise problems.

Root relief also is applied to eliminate interference between mating tooth profiles. The hob tooth is thicker near the top and removes more material near the root of the gear tooth.

A root fillet radius is always used in good gear design practice. The size of the radius can be increased to improve strength and durability of heavy-loaded gears.

Tool Material

Solid hobs and inserted blade hobs are made from high-speed steel, hardened to Rockwell C 64 to 66. In addition, low-temperature surface hardening processes--such as nitriding--are applied to increase the wear resistance and durability of high-speed steel hobs.

Recent developments in tool materials and coatings have dramatically increased achievable speeds and feeds in hobbing operations. The introduction of Titanium Nitride coatings in the early 1980s has tremendously boosted production rates in metal removal processes.

Titanium Nitride (TiN) coating, as a means of improving tool life, was first applied to gear cutting tools in 1980 in Japan. Typical TiN coatings have a hardness of approximately 80 to 85 Rockwell C and a thickness of only 0.00012 in. (0.003 mm). The extremely thin layer of coating makes it ideally suited for use on very precise tools like hobs and shaper cutters.

The high hardness of TiN coating, 85 Rockwell C versus 65 Rockwell C for high-speed steel, provides excellent wear and abrasion resistance to heavy and prolonged cutting loads. The high hardness enhances tool life. TiN has a lower coefficient of friction than high-speed steel. This reduces the cutting forces and heat generated during cutting. It also allows higher feeds and speeds. Actual tests have shown an increase of 30% to 50% in cutting speed and 10% to 20% in cutting feed without loss of tool life.

Applications of carbide hobs have increased in recent years with the introduction of rigid heavy-duty hobbing machines. Carbide hobbing can only be handled by modern machines. This limitation is caused by backlash and lack of stiffness. In addition, vibrations have a detrimental effect on tool life. CNC hobbers, with their independent servo motors for hob slide and the work table, are ideally suited for carbide hobbing. Especially important are the absence of play on the hob slide, stiffness of the work clamping fixture, and axial and radial runout on the work table.

There are three different applications of carbide hobbing.

1. Hobbing of soft gears with solid carbide hobs.
2. Hobbing of soft gears with indexable insert hobs.
3. Hobbing of hard gears.

Solid carbide hobs for soft gears are used in high-volume gear lines where the higher feeds and speeds attainable with carbide, and the resulting shorter cycle times, have an impact on the number of machines required for the operation.

One important factor to consider is the cost of carbide hobs. This cost can be 10 to 20 times higher than high-speed steel.

Hobs with indexable inserts are applied on a limited scale for roughing of gears with diametral pitch larger than 3. Indexable insert hobs consist of a cutter body made of ordinary tool steel equipped with replaceable carbide inserts. The cutter is mounted on a short arbor for maximum stiffness and rigidity. The main advantages of indexable insert hobs are:

* Carbide inserts are economical and permit full exploitation of carbide feeds and speeds;
* Life of the cutter body is virtually unlimited; and
* Regrinding is not required.

Finishing of gears after hardening also is possible with indexable insert hobs.

Carbide hobbing, or skiving, of gears after heat treatment is an alternative to gear grinding. This process combines the accuracy of grinding with the productivity of hobbing. Surface finish of the gear flanks is not comparable with grinding quality due to the typical tool marks left by the hob. Honing is recommended as a surface finish improvement operation after hard hobbing.

Skive hobbing is performed with a solid carbide hob with a negative rake angle. Material is only removed from the tooth flanks. The root is left untouched by the carbide hob. Therefore it is necessary to rough cut the gears prior to heat treatment with a protuberance hob, which leaves the desired machining stock on the flanks only.

Carbide hobs are resharpened with a diamond grinding wheel preferably using the deep grinding method which removes the entire grinding stock in one slow pass of the grinding wheel.

SHAPING

Gear shaping, according to the generating process developed by Fellows Corporation, operates on the principle of two gears rolling in mesh. The work blank, representing one of the two gears rotates around its axis, while the cutter, representing the other gear, reciprocates across the face of the work blank.

The rate of stroking and the length of stroke of the cutter determine the cutting speed. The rate of work rotation determines the generating feed. Just like in hobbing, the involute profile is generated in small increments by successive cuts of the cutter through the workpiece as they rotate together. These increments are illustrated in Figure 1-12.

Since the shaper cuts only in one direction, a relief action is provided for the return stroke. This is usually done by separating cutter and work. In some gear shaping machines, the work table moves away from the cutter on the return stroke. On other machines, the relieving mechanism is built into the cutter head. The latter design allows faster cutting speeds because less mass must be moved.

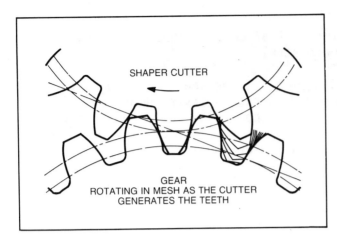

Figure 1-12. Generating action of a shaper cutter. (3)

The radial infeed of the cutter to full cutting depth is controlled by a mechanical cam on conventional shapers. A mechanical roller, connected to the cutter spindle, follows the contour of the cam. It gradually brings the cutter to full depth. Cams are available in various configurations for one cut, two cuts, three cuts, and four cuts cycles.

The gear shaper with its associated tooling accurately produces both external and internal spur, helical gears, and splines. Gears may be cut not only on conventional blanks but against flanges and in blanks where only narrow recesses are provided for cutter clearances. Certain types of gears can be generated by the shaper method only. These include: cluster gears, shoulder gears, solid herringbone gears, and internal gears.

Shaping also can be used for gears that require an exceptionally high surface finish or accuracy. This results because (in shaping) the stroking speed is independent from the generating feed. By increasing the stroking speed, the profile is generated by a larger number of flats resulting in finer finish.

Gear Shaping Machines

While gear shapers have traditionally been regarded as somewhat slower than hobbing machines, recent machine designs incorporate innovations providing operating speeds to 2100 strokes per minute. CNC shaping machines have greatly increased the versatility of the process. At the same time, they have improved productivity by eliminating idle times for parts requiring multiple setups. Spindle speeds and generating feeds, selected by change gears on conventional gear shapers, are infinitely variable on CNC shapers.

Gear shapers are built in various sizes and capacities to satisfy the needs of manufacturing. The smallest machine used in traditional gear production for automotive, agricultural, and earth-moving equipment can cut gears to 10 in. (254 mm) in pitch diameter and 2 in. (50.8 mm) maximum width. Larger machines can cut internal gears up to 106 in. (2692.4 mm) pitch diameter and 16 in. (406.2 mm) face width. For high-production applications, gear shapers can be automated with automatic part loaders.

(3) Drozda, Thomas J, Wick, Charles. Tool and Manufacturing Engineers Handbook, Fourth Edition, Volume 1: Machining. Dearborn, Michigan: Society of Manufacturing Engineers, 1983, p. 13-64.

Shaper cutters are just like hobs versatile tools with a wide range of application. For a given shaper cutter, the number of teeth on the gears to be cut can be varied within limits provided that the pitch, helix angle and pressure angle are the same.

Shaper Cutters

Shaper cutters are made in three basic styles (Figure 1-13).
* Disc-type,
* Deep counterbore-type, and
* Shank type.

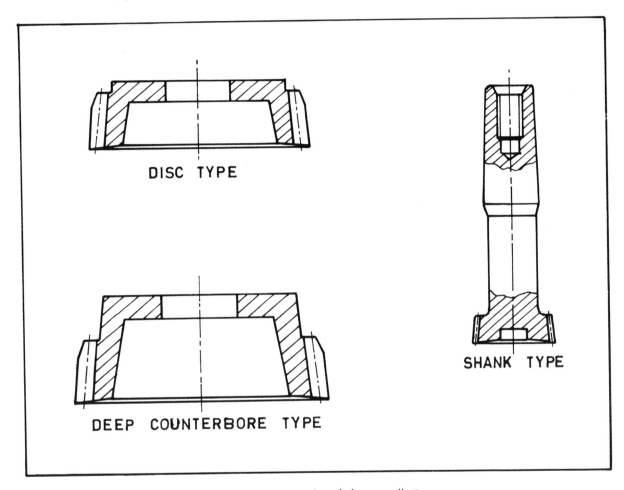

Figure 1-13. Basic styles of shaper cutters.

Disc-type shaper cutters are the most commonly used for external gears and splines. Disc cutters are installed directly on the machine spindle. Deep counterbore cutters are made from a thicker blank and have a deep recess in which the holding nut or screw can be concealed. This cutter cuts gears with a close shoulder, or internal gears. Taper shank cutters normally cut small internal gears or splines. Taper shank cutters are made in various taper sizes. The taper fits directly in the machine spindle, or indirectly with a reduction taper bushing. To minimize deflection, the pitch diameter of the cutter should approximate the diameter of the taper.

Commercial quality shaper cutters for spur gears with 20 degrees pressure angle are usually available from stock. High accuracy precision ground cutters for finishing operations, and ground cutters for pre-shaving, pre-grinding, and roughing operations are made per special order. Shaper cutters are made in two quality classes, class A cutters for finishing and pre-shaving and class B for roughing and pre-grinding.

Standard shaper cutters are designed to produce a true involute without modifications to the tooth profile. Often gear design mandates the use of modified profiles to improve performance. Shaper cutters can be ordered with various profile modifications (Figure 1-14).

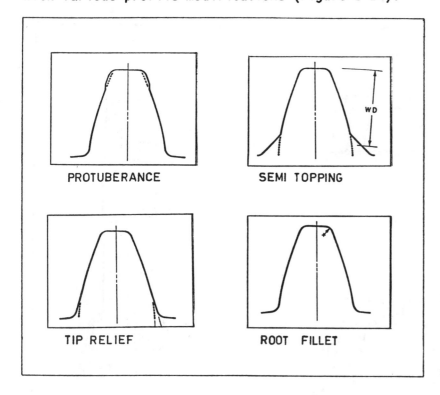

Figure 1-14. Various modifications to shaper cutter profile.

Protuberance cutters have a profile which is thicker near the tip to provide an undercut in the root of the gear teeth. The undercut in the gear root area acts as a clearance. The undercut is intended to facilitate subsequent finishing operations. Protuberance cutters are frequently applied for gears which are finished by shaving or grinding.

The root area of semi-topping cutters is modified at an angle to create a chamfer at the tip of the generated gear teeth. This feature is very helpful in minimizing damage due to nicks and burrs during part handling. These nicks and burrs could result in gear noise. Tip relief cutters remove slightly more material near the top of the gear tooth. Tip relief compensates for interference between mating profiles and usually results in reduced gear noise.

Root fillet radius modification consists of blending the corners of the tool tip with a radius which produces a fillet in the root corners of the generated teeth. A root fillet increases bending strength of the gear teeth and improves fatigue life. Tool wear on the shaper cutter also is improved, as a sharp corner would chip off very quickly.

Helical gears are cut with a helical shaper cutter. The cutter spindle is rotated during the reciprocating motion by means of a special helical cam or guide. Each helix angle requires its own special helical guide. A helical shaper cutter and its guide form a set of tooling required to cut a specific helical gear. Two helical cutters and two guides are required to cut a pair of meshing helical gears (left and right hand gears) (Figure 1-15).

shaper cutter	external gear	internal gear
Right hand	Left hand	Right hand
Left hand	Right hand	Left hand

Figure 1-15. Relationship between hand of shaper cutter and helical gear.

Helical shaper cutters require generally a wider recess at the bottom of the cut for full cutter runout.

Shaper cutters are made of high-speed steels containing molybdenum and tungsten. They are heat treated to a Rockwell C hardness of 64 to 65. A nitriding surface treatment improves wear resistance further.

Titanium nitride coating often is applied to increase permissible feeds and speeds on modern shaping machines. TiN coated shaper cutters increase productivity by as much as 30% without decrease in tool life.

Carbide shaper cutters are rarely used in normal gear production. The advantages of carbide tools cannot be fully exploited on gear shapers because the machines are limited in the maximum reciprocating speed of the cutter spindle. As opposed to hobbing, where the hob material limits the permissible cutting speed, the limiting factor on shapers is the maximum speed of the reciprocating mechanism of the cutter spindle.

SHAVING

Shaving is strictly a finishing method for soft gears which have been rough cut by hobbing or shaping. The shaving cutter is a helical gear which rotates without backlash with the work gear. The helix angle of the shaving cutter is different from the helix angle of the gear. Cutter and gear rotate, therefore, in different planes. The angle between the axes of cutter and gear is called the "crossed axis angle." The crossed axis angle is the algebraic sum of the cutter helix angle and the gear helix angle. Examples of crossed axis angles are listed in Table 1-1.

Table 1-1
Examples of Crossed Axis Angles

Cutter helix	Gear helix	Crossed axis angle
10 degr R.H.	22 degr L.H.	+10 -22 = -12 degr
15 degr R.H.	0 degr SPUR	+15 -0 = +15 degr
12 degr L.H.	25 degr R.H.	-12 +25 = +13 degr

The + or - sign before the crossed axis angle indicates the direction in which the cutter head must be swivelled. A crossed axis angle between 10 and 15 degrees is recommended for good shaving results.

Because of the crossed axes, a sliding action occurs between the tooth profiles of cutter and gear, and cutting action occurs along the cutting edges of the cutter. Cutting edges in the shaving cutter are formed by parallel serrations which run along the face width of the cutter. During the shaving cycle, cutter and work gear are pressed against each other by feeding the work vertically into the cutter. Traversing feed motion is accomplished by longitudinal travel of the work piece across the face of the cutter. At the end of each stroke, in-feed is applied by moving cutter and work piece closer together. The cutter rotation is reversed and the traverse feed direction is reversed. The sequence of feed strokes is repeated until the machining stock is completely removed. The normal shaving cycle consists of four passes with vertical upfeed and two passes without upfeed.

The amount of upfeed at each stroke is set by an upfeed cam. The cam is flexible and can be adjusted for diminishing increments. For instance, the feed cam can be set so that the first upfeed is 0.003 in. (0.08 mm), the second 0.002 in. (0.05 mm), the third 0.001 in. (0.03 mm), the fourth 0.0005 in. (0.013 mm), and two strokes with zero upfeed.

There are four different methods of shaving: conventional shaving, diagonal shaving, underpass shaving, and plunge shaving. The working principle of each method will be discussed in more detail in a later chapter.

The difference between the four methods lies in the direction of reciprocation of the work gear relative to its own axis. In conventional shaving, the reciprocation is parallel to the work axis. In diagonal shaving, the direction of reciprocation is at an angle with the work axis. In underpass shaving, the direction of reciprocation is perpendicular to the work axis. Plunge shaving has no reciprocation. Plunge shaving is accomplished by moving the centers of cutter and work gear closer together in one continuous feed motion.

Conventional shaving is the slowest method. But it is also the most universal and employs an inexpensive shaving cutter design. Diagonal shaving and underpass shaving are faster methods. But they require specially designed cutters with wider face width. These cutters are more expensive and often more complex to regrind.

Since hobbing and shaving operations follow close upon each other in gear machine lines, the speed of the shaving operation often is of no consequence to the output of the line. The speed at which parts move through the line is determined by inherently slower operations, like hobbing or shaping. Therefore, the conventional shaving method will, in many cases, be selected because of more economic tooling.

The proximity of a shoulder may pose difficulties in finishing a gear with conventional shaving because of interference between cutter and shoulder. Diagonal shaving or underpass shaving are the indicated solutions for finishing shoulder gears.

During shaving, the work is mounted between centers and is driven by the shaving cutter. Arbors locate and clamp gears between centers (Figure 1-16). Shafts are mounted directly between the machine centers (Figure 1-17).

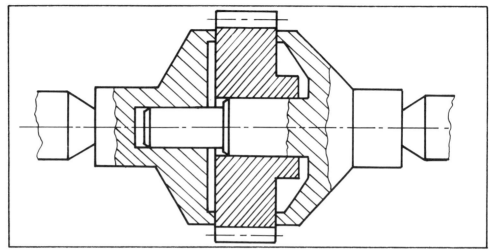

Figure 1-16. A solution for loading parts on a shaving machine.

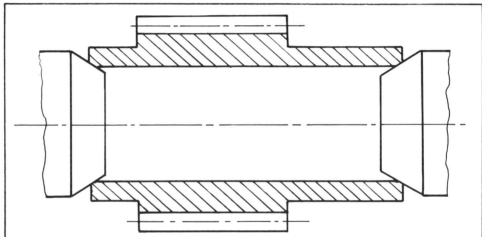

Figure 1-17. A second solution for loading parts on a shaving machine.

In the true sense of the word, shaving is not a generating method because of the absence of a gear train that links the rotation of the cutter to the indexing speed of the gear. The work gear rotates freely between centers and is driven by the cutter. Since the amount of stock removed in the shaving operation is small, the amount of quality improvement that can be expected also is limited. Shaving improves considerably the involute profile and surface finish. Tooth spacing errors, eccentricity, and helix angle errors also can be corrected but to a lesser extent. Therefore the pre-shave gear quality is of the utmost importance and determines the amount of quality improvement that can be obtained by shaving. Normally, the quality of hobbed and shaped gears can be improved at least two quality classes by the shaving process.

The shaving machine has the capability to longitudinally crown the gear teeth by sinking the cutter deeper into the work near the ends of the gear. Crowning action is obtained by rocking the table around a central pivot point. The rocking motion combined with the traversing stroke of the work table results in reduced distance between cutter and gear, and therefore deeper working depth of the cutter.

The slightly thinner tooth at the ends ensures that the load between mating gears is transmitted near the center of the face width. End loading of the gear teeth and undesirable axial thrust loads on bearings are avoided by crowning.

The crowning feature is used with the conventional shaving method, and when shaving diagonally with traverse angles smaller than 50 degrees. With higher traverse angles, the full amount of crown may not be obtained in diagonal shaving because of the short stroke length. In some cases, when it is impossible to obtain adequate crowning of the gear teeth by the rocking motion of the table (in diagonal shaving or underpass shaving), the crowning can be ground in the cutter lead. A shaving cutter with a hollow helix will produce a gear with longitudinal crowning (Figure 1-18).

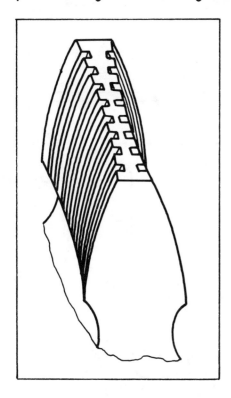

Figure 1-18. Shaving cutter with hollow lead for crowning of gear tooth.

Also the involute profile can be modified by grinding a hollow involute (Figure 1-19). The cutter will then remove slightly more material near the top and root of the gear tooth. These involute modifications were previously defined as tip and root relief. Shaving cutters with lead and involute modifications are often used to compensate for heat treat distortions.

Shaving Machines

Shaving machines are built in various types and sizes to handle a range of gears from .5 to 180 in. (12.7 to 457.2 mm) pitch diameter. The most frequently used machine is the knee-type shaver. It can handle gears to 18 in. (457.2 mm) pitch diameter, diametral pitch 3 in. (76.2 mm), and with a maximum table travel of 6 in. (152.4 mm). In a knee-type shaving machine, the work gear is mounted between centers with its axis horizontal. This type of machine is easy automated and adapted to high-volume production lines.

Larger shaving machines usually are designed to take the work piece with the axis in a vertical position. The vertical machine configuration is more rigid and better suitable to accommodate heavy work pieces.

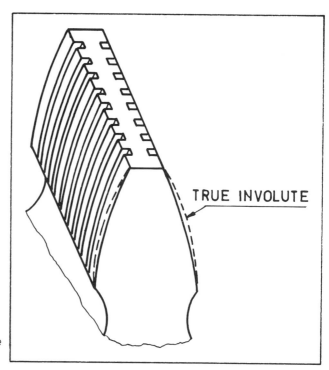

Figure 1-19. Shaving cutter with modified involute profile.

Special machines have been designed to shave exclusively internal gears. Internal shavers are normally assigned to dedicated operations for production of large series of internal gears. External gear shavers can also be equipped with a special internal cutter head to permit shaving of internal gears in small lots.

Shaving Cutters

Shaving cutters are made of superior grade high-speed steel, heat treated to 64 - 66 Rockwell C, and precision ground to very accurate tolerances. Two quality classes are available, AA and A, AA being used for the most precise work. Shaving cutters are made for gears with diametral pitch from 4 to 64, the pitch diameter of the cutter ranges normally between 2.5 and 12.5 in. (63.5 to 317.5 mm). Standard cutter widths for conventional shaving range from .5 to 1.0 in. (12.7 to 25.4 mm). Shaving cutters are very accurate tools. These tools must be handled and stored with proper care to ensure a long service life.

2
Hobbing

THE PROCESS

The engagement of hob and workgear can be compared with the interaction of worm and wormwheel. The hob is like a worm with cutting edges, which is fed axially across the face of the workgear while it rotates in constant mesh with the gear.

For each revolution of the hob, the workgear advances one tooth. A gear with a number of teeth equal to Z is rotated over $1/Z$ of a full revolution for each revolution of the hob. If the hob has a cutting speed of N revolutions per minute, then the corresponding rpm of the workgear is:

$$N_{work} \times Z_{work} = N_{hob} \times Z_{hob}$$

$$N_{work} = \frac{N_{hob} \times 1}{Z_{work}}$$

If the hob is advanced axially over the face width of the gear with a feed of F inches per revolution of workgear, then the feedrate in inches per minute is equal to:

$$\text{Feedrate} = N_{work} \times F$$

The time required to complete one cut can be calculated by dividing the length of cut L by the feedrate:

$$T = \frac{\text{Length of cut}}{\text{Feedrate}}$$

$$T = \frac{L}{N_{work} \times F}$$

$$T = \frac{Z_{work} \times L}{N_{hob} \times 1 \times F}$$

A worm with two threads rotates the wormwheel over two teeth for each revolution of the worm. Similarly, a hob with two threads will index the gear two teeth for each revolution of the hob. If K represents the number of threads in the hob then the cutting time is:

$$T = \frac{Z_{work} \times L}{N_{hob} \times K \times F}$$

Cycle Time Formulas

Depending on the tooth size, gears and splines are hobbed in a single pass, or in a two pass cycle consisting of a roughing cut followed by a finishing cut. State-of-the-art hobbing machines have the capability to vary cutting parameters between first and second cut so that a different formula calculates cycle times for single cut and double cut hobbing.

Single Cut Hobbing Cycle

The cycle time is given by the equation:

$$T = \frac{Z \times L}{N \times K \times F}$$

where: T = cycle time in minutes
Z = number of gear teeth
L = length of cut in inches
N = hob revolutions per minute
K = number of hob starts
F = feedrate in inches per revolution of work

Double Cut Hobbing Cycle

The cycle time is given by the equation:

$$T = \frac{Z \times L1}{N1 \times K \times F1} + \frac{Z \times L2}{N2 \times K \times F2}$$

where: T = hobbing time in minutes
Z = number of gear teeth
$L1$ = hob travel in inches first cut
$L2$ = hob travel in inches second cut
$N1$ = hob revolutions per minute first cut
$N2$ = hob revolutions per minute second cut
K = number of hob starts
$F1$ = feedrate in inches per revolution of workgear, first cut.
$F2$ = feedrate in inches per revolution of workgear, second cut

Some of the parameters of the cycle time formulas, like the number of teeth of the gear, can be found directly on the part print. Others require additional calculations before they can be entered in the equation.

It is important to know that diametral pitch and pitch diameter of the workgear determine the size of the hobbing machine required for the job. The size of the gear tooth will also influence the feedrate used to cut the gear, and whether the gear must be hobbed in a single or double cut cycle.

CALCULATION OF HOB TRAVEL (L)

The hob travel length consists of the following four elements:

* Gear face width.
* Spacer width.
* Hob approach.
* Hob overrun.

Gear Face Width

The gear face width is also indicated on the part print as the width of the gear blank. When more than one part is loaded per cycle, the total gear width must be taken into account (Figure 2-1).

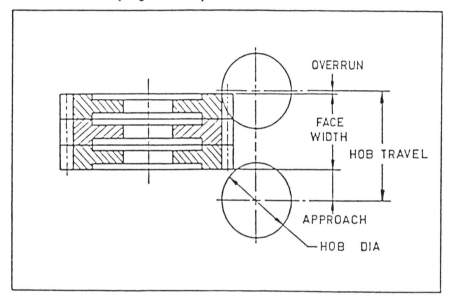

Figure 2-1. Elements of hob travel.

The advantage of loading more than one part per cycle is that the approach and overrun length per part is reduced proportionally with the number of parts in the stack. With the reduction in length of cut goes an equivalent reduction in machining time as illustrated in Table 2-1.

Table 2-1
Length of Cut and Machining Time

	1 part	2 parts	3 parts
Gear face width	1.215	2.430	3.645
Hob approach	1.068	1.068	1.068
Hob overrun	0.021	0.021	0.021
Total length of cut	2.304	3.519	4.734
Length of cut per part	2.304	1.759	1.578
Cutting time per part	4.179 min.	3.191 min.	2.862 min.
Reduction in time	-	24%	32%

Spacer Width

Gear configuration may be such that a spacer is required between gears to load more than one part per cycle. In this case the width of the spacer must be added to the total face width (Figure 2-2). Obviously, the spacer width must be smaller than the sum of approach and overrun in order to take advantage of stacking two parts per cycle.

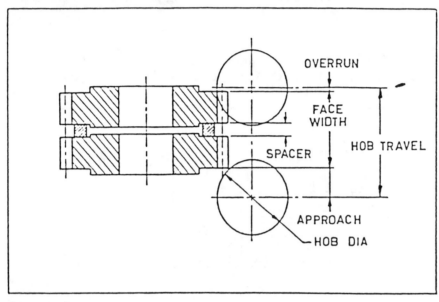

Figure 2-2. Use of a spacer to stack gears.

Approach

Hob approach is the distance from the point of initial contact between hob and gear blank to the point where the hob reaches full depth of cut. The approach length is a function of hob diameter, gear outside diameter, depth of cut, and gear helix angle.

The hob approach length is calculated with the formula:

$$A = \sqrt{W \times \left[\frac{D + G - W}{\cos^2(H)} - G \right]}$$

where: A = hob approach length in inches
W = depth of cut in inches
D = hob outside diameter in inches
G = gear outside diameter in inches
H = gear helix angle

For spur gears (as seen in Figure 2-3) $H = 0$ and $\cos H = 1$, and the approach formula is simplified to:

$$A = \sqrt{W \times (D - W)}$$

In a single cut cycle, the depth of cut is:

$$W = \frac{\text{Gear outside dia} - \text{Gear root dia}}{2}$$

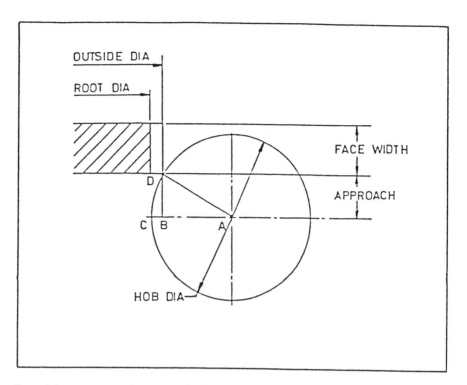

Figure 2-3. Hob approach for spur gears.

In a double cut cycle, the approach travel for roughing is longer than for finishing because of the difference in cutting depth (Figure 2-4).

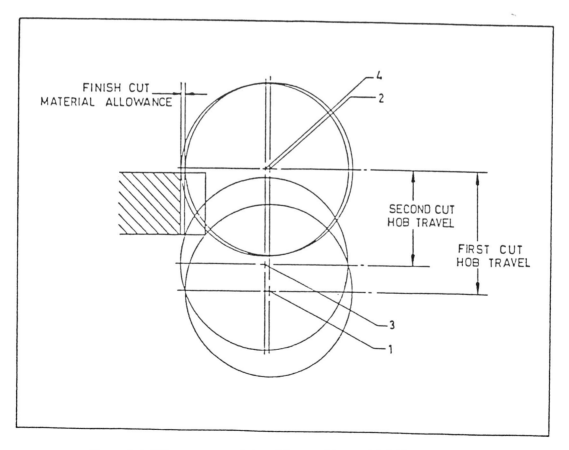

Figure 2-4. Different approach travel for roughing and finishing.

Overrun

Hob overrun is the linear hob travel beyond full cutting depth, required to complete generation of the gear teeth.

Hob overrun is calculated with the formula:

$$R = \frac{S \times \cos(H) \times \tan(SA)}{\tan(PA)}$$

where: R = hob overrun in inches
S = addendum of gear tooth in inches
H = gear helix angle
SA = hob head swivel angle
PA = gear pressure angle

The hob head swivel angle is a function of helix angle and hand of both workgear and hob (see Table 2-2).

Table 2-2
Calculation of Hob Head Swivel Angle

Gear helix hand	Hob helix hand	Hob head swivel angle
left	left	H − HB
left	right	H + HB
right	left	H + HB
right	right	H − HB

In the table above, HB represents the hob helix angle. The minimum hob head swivel angle is obtained when the helix of gear and hob have the same hand.

All formulas are based on the theoretical points of contact between hob and work piece. In practice, clearance between hob and workgear is needed to assure safe cutting conditions. Therefore, a clearance amount of 0.040 to 0.100 in. (1.02 to 2.54 mm) must be added to the theoretical values of approach and overrun.

For spur gears $H = 0$ degrees, $\cos(H) = 1$ and $SA = HB$. The overrun for spur gears is then:

$$R = \frac{S \times \tan(HB)}{\tan(PA)}$$

For a seven diametral pitch gear, with 20 degrees pressure angle and hobbed with a three degree helix hob, the overrun is:

$$R = \frac{.1429 \times .05241}{.36397} = .020 \text{ inch}$$

It is obvious that for practical purposes the theoretical calculation of hob overrun for spur gears can be replaced by a fixed value which includes clearance, for instance of 0.100 in. (2.54 mm).

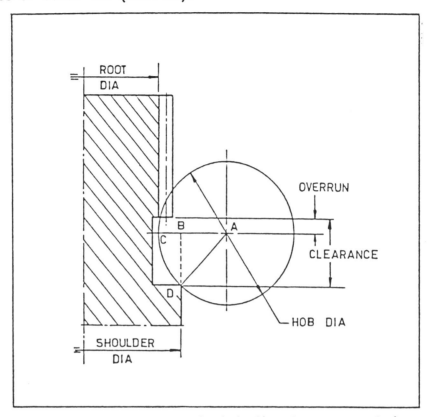

Figure 2-5. Hob interference with adjacent shoulder.

Special attention must be paid in cases where the hob diameter runs out in a shoulder adjacent to the gear. The minimum amount of clearance between gear and shoulder (see Figure 2-5) to avoid cutting marks in the shoulder can be calculated as follows:

$$CL = R + BD$$

where: CL = minimum clearance
R = hob overrun
BD = distance between point of contact and center of hob

$$BD^2 = AD^2 - (AC - CB)^2$$

$$AC = AD = 1/2 \text{ hob diameter} = D/2$$

$$CB = SD/2 - RD/2$$

SD = shoulder diameter
RD = root diameter

$$CL = R + \sqrt{(D/2)^2 - (D/2 - SD/2 + RD/2)^2}$$

$$= R + 1/2 \sqrt{D^2 - (D - SD + RD)^2}$$

Maximum clearance is required when $(D - SD + RD) = 0$ or when the shoulder extends to the centerline of the hob.

$$CL_{max} = R + 1/2D$$

HOB REVOLUTIONS PER MINUTES (N)

Cutting speed in a hobbing operation is defined as the peripheral velocity of the hob.

$$V = \frac{\pi \times D \times N}{12}$$

where: V = Cutting speed in surface feet per minute (SFPM)
  D = Hob diameter in inches
  N = Revolutions per minute of the hob
  π = 3.14159...

Hobs are produced and stocked in standard sizes with specific dimensions per diametral pitch range. These standard hobs are mainly used for low-volume production and job shop work. High-volume gear manufacturers specify their own hob dimensions to suit each individual application. Hob diameter values (in Table 2-3) can be used as reference when estimating cycle times. Multiple thread hobs require special design consideration and usually have a larger outside diameter than a comparable single thread hob.

Table 2-3
Standard Hob Sizes for Single Thread Hobs

Diametral Pitch	Hob diameter
DP < 12.7	2.50
12.7 < DP < 8	3.00
8 < DP < 7	3.25
7 < DP < 5	3.50
5 < DP < 4	4.00
4 < DP < 3	4.50
3 < DP < 2.5	5.00
2.5 < DP < 2	5.50

To find the number of revolutions of the hob the formula is converted to:

$$N = \frac{12 \times V}{\pi \times D}$$

As in most metalcutting processes, there are no specific values of speeds and feeds that must be used. Cutting paramaters are in fact dependent on many variables and starting values are often determined by past experience.

Speeds and feeds in a hobbing operation are affected by:

* Physical properties of tool materials;
* Machinability of work material;
* Quality specifications;
* Rigidity of machine and fixture;
* Desired tool life, and
* Cutting fluids, lubricants and coolants.

The three most important factors which affect cutting speed are the hob material, its treatments and coatings, and the machinability of the work gear material.

Table 2-4
Cutting Speeds for High-speed Steel Hobs

Material	Brinell Hardness	Cutting speed	
		Rough feet/min	Finish feet/min
SAE 8615	161 - 205	185 - 235	270 - 295
SAE 8620	170 - 217	180 - 220	260 - 290
SAE 8630	179 - 229	170 - 200	240 - 275
SAE 1015	131 - 170	170 - 200	240 - 275
SAE 1045	179 - 229	145 - 175	205 - 230
SAE 1065	183 - 241	120 - 150	175 - 200
SAE 5140	174 - 229	160 - 190	215 - 245
SAE 4135	187 - 229	170 - 200	240 - 265
SAE 4130	187 - 229	180 - 220	260 - 290

Table 2-5
Cutting Feeds for Single Thread High-speed Steel Hobs

Material	Brinell Hardness	generating feed in inch/revolution		
		Diametral pitch		
		25 - 10	10 - 4	4 - 2
SAE 8615	161 - 205	0.100 - 0.150	0.120 - 0.180	0.080 - 0.100
SAE 8620	170 - 217	0.100 - 0.150	0.120 - 0.180	0.080 - 0.100
SAE 8630	179 - 229	0.100 - 0.140	0.120 - 0.160	0.080 - 0.100
SAE 1015	131 - 170	0.090 - 0.140	0.120 - 0.160	0.070 - 0.100
SAE 1045	179 - 229	0.080 - 0.120	0.100 - 0.140	0.060 - 0.080
SAE 1065	183 - 241	0.080 - 0.120	0.100 - 0.140	0.060 - 0.080
SAE 5140	174 - 229	0.100 - 0.130	0.120 - 0.160	0.080 - 0.100
SAE 4135	187 - 229	0.100 - 0.130	0.100 - 0.140	0.080 - 0.100
SAE 4130	187 - 229	0.100 - 0.130	0.100 - 1.140	0.080 - 0.100

The values listed in the preceding table represent average feeds for single cut hobbing. The feedrate can be increased for finishing with 30 to 40%. Titanium nitride coated hobs permit cutting speeds which are 30% higher than high speed steel. Cutting speeds and productivity can be further increased with solid carbide hobs.

Number of Hob Starts

Number of hob starts and cycle time are inversely related to each other. Cycle time decreases when the number of hob starts is increased.

A single start hob rotates the work one tooth for each revolution of the hob. With a 2,3,4 start hob the work is rotated over 2,3,4 teeth for each revolution of the hob. Assuming the same feedrate for multistart hobbing as for single start, the cycle will be completed 2,3,4 times faster.

Quality considerations limit the application of multistart hobs, for the following reasons:

* Fewer hob teeth participate in the generation of the tooth profile, it is therefore less accurate.
* Multistart hobs have an inherent thread spacing error which is repeated in the workpiece under certain conditions.

The following guidelines should be followed when estimating times with multistart hobs:

* The number of teeth in the gear must not be divisible by the number of hobstarts.
* Only use multistart hobs for gears with a large number of teeth. ($Z > 25$)

When working with multistart hobs the feedrate must be reduced to compensate for the increased toothloading of the hob. The following reduction factors in Table 2-6 are recommended.

Table 2-6
Feedrate Reduction Factors for Multi-start Hobs

Number of hobstarts	Reduction factor
1	1
2	0.75
3	0.65
4	0.60

Example: Normal feedrate with single start hob is 0.16 in. (4.1 mm) per revolution of workpiece. When using a two start hob for the same job the feedrate should be reduced to:

$$.75 \times .160 = .120 \text{ in/revolution}$$

EXAMPLES OF CYCLE CALCULATIONS

Example 1

Transmission gear hobbed on arbor fixture (Figure 2-6).

Part print data

number of teeth	61
diametral pitch	7
pitch diameter	8.714
outside diameter max	8.990
outside diameter min	8.985
root diameter max	8.346
root diameter min	8.336
pressure angle	20 degrees
helix angle	0
face width	1.215
material	SAE 8620

Hob data

outside diameter	4.60
number of starts	1
spiral angle	4.25 degrees
material	HSS

Machine setting data

double cut cycle	
cutting speed rough	230 sfpm
cutting speed finish	290 sfpm
feedrate rough	0.177 ipr
feedrate finish	0.236 ipr
number of parts per cycle	2
spacer width	0.260
finish cut material allowance	0.060

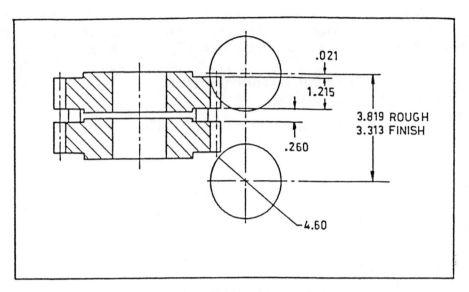

Figure 2-6. Example 1: Hobbing of transmission gear.

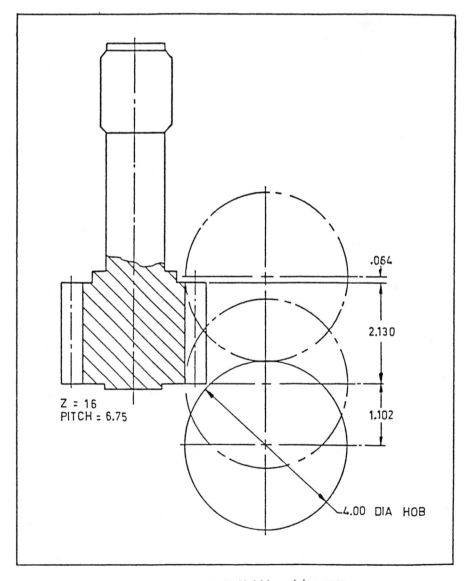

Figure 2-7. Example 3: Hobbing pinion gear.

Cycle Time Calculation

$$\text{Hob rpm rough} = \frac{12 \times 230}{3.14159 \times 4.6} = 191 \text{ rpm}$$

$$\text{Hob rpm finish} = \frac{12 \times 290}{3.14159 \times 4.6} = 240 \text{ rpm}$$

$$\text{Gear addendum} = \frac{8.9875 - 8.7142}{2} = .137$$

$$\text{Whole tooth depth} = \frac{8.9875 - 8.341}{2} = .323$$

Depth of cut roughing cut = .323 − .060 = .263

Depth of cut finishing cut = .060

Hob approach roughing cut = $\sqrt{.263 \times (4.60 - .263)}$ = 1.068

Add .040 clearance = .040 + 1.068 = 1.108

Hob approach finishing cut = $\sqrt{.060 \times (4.60 - .060)}$ = .522

Add .040 clearance = .040 + .522 = .562

$$\text{Hob overrun rough and finish} = \frac{.137 \times \cos 0 \times \tan 3.25}{\tan 20}$$

$$= \frac{.137 \times .05678}{.36397} = .021$$

Add .040 clearance = .040 + .021 = .061

Total hob travel roughing = 1.068 + .061 + 2.430 + .260 = 3.819

Total hob travel finishing = .562 + .061 + 2.430 + .260 = 3.313

$$\text{Cycle time} = \frac{61 \times 3.819}{190 \times .177} + \frac{61 \times 3.313}{240 \times .236}$$

$$= 10.495 \text{ min for 2 pieces}$$

$$= 5.297 \text{ min for 1 piece}$$

Example 2

Same as Example one, except that the gear is now hobbed with a two start hob. We will assume that the two start hob has the same outside diameter as the single start hob so that approach and overrun values are the same as in the previous example.

$$\text{Feedrate for roughing} = .75 \times .177 = .133$$

$$\text{Feedrate for finishing} = .75 \times .236 = .177$$

$$\text{Cycle time} = \frac{61 \times 3.819}{190 \times 2 \times .133} + \frac{61 \times 3.313}{240 \times 2 \times .177}$$

$$= 6.987 \text{ min for 2 pieces}$$
$$= 3.493 \text{ min for 1 piece}$$

The savings in cycle time is

$$10.495 - 6.987 = 3.508 \text{ min} = 33.4\%$$

This example illustrates clearly the increased productivity which results from the use of multistart hobs.

Example 3

CNC hobbing applications use low feedrate for the approach length and high feedrate for the full depth cut. The objective here is reducing cycle time by optimizing feedrates during the cut. The feedrate is changed during the cut by the CNC control (Figure 2-7).

Part print data

number of teeth	16
diametral pitch	6.75
pitch diameter	2.3703
outside diameter max.	2.760
outside diameter min.	2.754
root diameter max.	2.1574
root diameter min.	2.1374
pressure angle	22.5000
helix angle	0
face width	2.130
material	SAE 8620

Hob data

outside diameter	4.000
number of starts	1
spiral angle	3.000
material	HSS

Machine setting - conventional hobber

single cut cycle	
cutting speed	190 sfpm
cutting feed	0.160 ipr
number of parts per cycle	1

Machine setting - CNC hobber

single cut cycle	
cutting speed	190 sfpm
cutting feed approach cut	0.160 ipr
cutting feed full depth cut	0.280 ipr
number of parts per cycle	1

Cycle time calculation - conventional hobber

$$\text{Hob rpm} = \frac{12 \times 190}{3.14159 \times 4} = 181 \text{ rpm}$$

$$\text{Gear addendum} = \frac{2.757 - 2.3703}{2} = .1933$$

$$\text{Depth of cut} = \frac{2.757 - 2.147}{2} = .305$$

$$\text{Hob approach} = \sqrt{.305 \times (4 - .305)} = 1.062$$

Add .040 clearance = .040 + 1.062 = 1.102

$$\text{Hob overrun} = \frac{.1933 \times \cos(1) \times \tan(3)}{\tan(22.5)} = .024$$

Add .040 clearance = .040 + .024 = .064

Total hob travel = 1.102 + 2.130 + .064 = 3.296

$$\text{Cycle time} = \frac{16 \times 3.296}{181 \times .160} = 1.821 \text{ min}$$

Cycle time calculation - CNC hobber

All the data calculated above can be used except for the total hob travel which is split between approach and full depth cut.

Approach length = 1.102

Full depth length = 2.130 + .064 = 2.194

$$\text{Cycle time} = \frac{16 \times 1.102}{181 \times .160} + \frac{16 \times 2.194}{181 \times .280}$$

$$= .609 + .693$$

$$= 1.302 \text{ min}$$

The reduction in cycle time is

$$1.821 - 1.302 = .519 \text{ min.} = 28.5\%$$

FIELD OF APPLICATION

Hobbing is the most widely used method of gear manufacturing. Its field of application, however, is restricted by part geometry. The major limitation is that hobbing is not applicable to manufacturing internal gears. Other gear cutting methods, like shaping, broaching, or skiving must be used.

Hobbing is also not applicable to shoulder gears because of the hob diameter runout in the shoulder. As indicated, the distance between gear face and an adjacent shoulder must be greater than the minimum required hob runout length. In some cases it is possible to cut shoulder gears by specifying hobs with reduced outside diameter. However, hob design considerations limit the variation in outside diameter.

Hobbing is without a doubt the most productive gear cutting method for external gears. It can be used as a semifinishing and finishing gear process. Hobbing as a finishing process is accomplished by rough and finish cutting the gears on the hobbing machine without a subsequent finishing operation. Most often hobbing is used in combination with a gear finishing operation like shaving or grinding.

Productivity can be increased by stacking several gears on the hobbing fixture. Stacks of more than two gears require good quality gear blanks with the gear faces parallel to each other and square to the bore.

One remarkable feature of the hobbing machine is the ability to make crowned or tapered gears. Crowning is often used in gear design practice to avoid end loading of the gear teeth. Taper hobbing can be used to compensate for uneven shrinkage in heat treatment.

Heat treated helical gears are typically affected by lead unwind, which is a change in helix angle after hardening. Lead angle variations are easily compensated for on a gear hobbing machine by installing sets of differential change gears, or by programming corrected helix angles on CNC controls.

The hobbing machine is capable of various cutting cycles.

Conventional Hobbing

The hob is fed down in the same direction as the hob rotation (Figure 2-8).

Cycle:

1 - 2 Rapid advance
2 - 3 Feed down
3 - 4 Rapid retract
4 - 1 Rapid return to start position

Climb Hobbing

The hob is fed up in opposite direction of the hob rotation (Figure 2-9). This method of hobbing is most often used because of smoother cutting conditions and more favorable chip forming.

Cycle:

1 - 2 Rapid advance
2 - 3 Feed up
3 - 4 Rapid retract
4 - 1 Rapid return to start position

Taper Hobbing

The center-distance between hob and work gear is reduced gradually during the feed stroke resulting in a tapered tooth form. The outside diameter of the gear can also be taper cut with a topping hob as shown in Figure 2-10. A non-topping hob will leave the outside diameter cylindrical. Taper hobbing compensates for heat treatment distortions or to compensate for bending deflections under heavy loads.

Cycle:

1 - 2 Rapid advance
2 - 3 Feed up while work gear feeds into the hob
3 - 4 Rapid retract
4 - 1 Rapid return to start position

Crown Hobbing

The hob can follow a curved path during the cutting stroke resulting in a crowned tooth form (Figure 2-11). Crown hobbing is used to avoid end loading of the gear teeth. By making the teeth slightly thinner at the ends, the contact pattern will always be located near the center of the gear face.

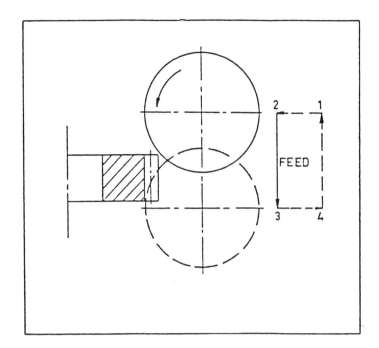

Figure 2-8. Conventional hobbing cycle.

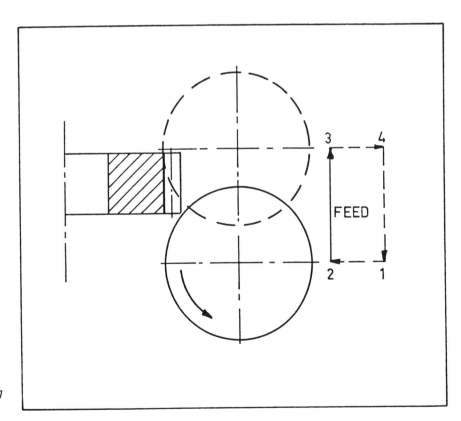

Figure 2-9. Climb hobbing cycle.

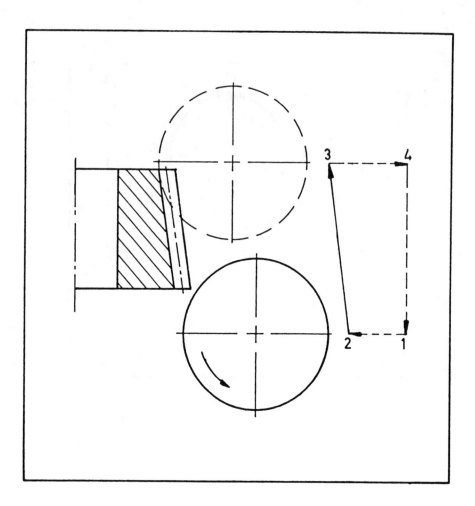

Figure 2-10. Taper hobbing cycle.

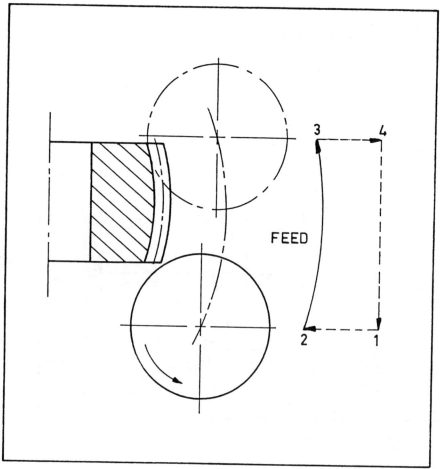

Figure 2-11. Crown hobbing cycle.

Cycle:

1 - 2 Rapid advance
2 - 3 Feed up while work gear feeds in and out.
3 - 4 Rapid retract
4 - 1 Rapid return to start position

Hob Two Gears with Same Diametral Pitch

Two gears with same diametral pitch and same pressure angle can be cut in the same cycle on a CNC hobber as shown in Figure 2-12.

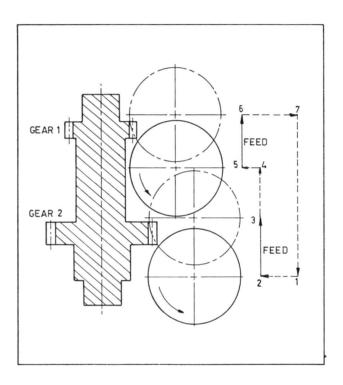

Figure 2-12. Hobbing of two gears in the same cycle.

Cycle:

1 - 2 Rapid advance
2 - 3 Feed through gear 2
3 - 4 Rapid advance
4 - 5 Rapid advance
5 - 6 Feed through gear 1
6 - 7 Rapid retract
7 - 1 Rapid return to start position

CNC hobbing with variable feed

Speeds and feeds can be changed during the cycle on a CNC hobbing machine. CNC hobbing with variable feed makes it possible to optimize the cutting cycle by adjusting the feedrate to the cutting conditions. The feedrate may be increased or decreased during various phases of the cycle to suit specific productivity and tool wear goals. (Figure 2-13)

Cycle:

1 - 2 Rapid advance
2 - 3 Approach cut at feedrate (1), progressively increasing cutting forces.
3 - 4 Full depth cut at feedrate (2), constant cutting forces.
4 - 5 Exit cut at feedrate (3), progressively decreasing cutting forces.
5 - 6 Rapid retract
6 - 1 Rapid return to start position

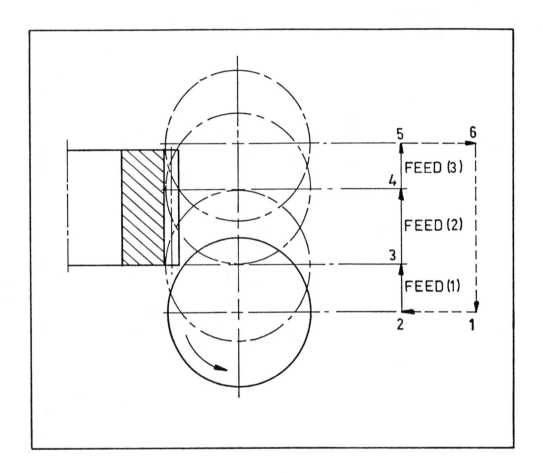

Figure 2-13. CNC hobbing cycle with variable feed rate.

3
Shaping

THE PROCESS

Shaping is based on the principle of two rotating gears which mesh with each other while one gear--the cutter--reciprocates across the face of the other gear. Cutting action occurs during the down stroke. The cutter is relieved by a cam during the upstroke to prevent rubbing of the cutting edge along the gear face.

The cutter is fed gradually to full cutting depth by reducing the center distance between cutter and workgear. Once the cutter has reached the cutting depth, a generating feed motion is engaged which finishes the shaping cut in one revolution of the workpiece.

CYCLE TIME ELEMENTS

The time required to complete one full cut on a gear shaping machine consists of two elements:

* Infeed time, and
* Rotary feed time.

The infeed time is the time required to feed the shaping cutter to cutting depth, starting from a position where the cutter clears the workpiece. Infeed is accomplished by moving the axes of cutter and work closer together by means of hydraulic controls, a mechanical cam, or CNC controls.

$$Ti = \frac{D + C}{Fi \times N}$$

where: Ti = infeed time in minutes
D = depth of cut in inches
Fi = infeed in inches/stroke
N = number of cutter strokes per minute
C = cutter clearance, usually .020 to .040 inch.

Very low feedrate values, in the range of 0.001-0.002 in. (0.03-0.05 mm) per stroke, are used to feed the shaper cutter to full depth.

The rotary feed time is the time required to complete one full generating cut. Once the cutter is fed into the work, infeed is stopped and a generating cut is taken until all teeth are cut to the same depth. Rotary feed is accomplished in one rotation of the workpiece. If more than one cut is required, the sequence of infeed and rotary feed is repeated until the tooth form is completely finished.

$$Tr = \frac{PCD \times \pi}{Fg \times N}$$

where: Tr = rotary feed time in minutes
PCD = pitch circle diameter of gear in inches
Fg = generating feedrate in inches/stroke
N = number of strokes per minute

Cutting Speed

The speed of the shaping cutter varies during the stroke, it is zero at upper and lower limits and reaches a maximum somewhere between those two points. The average reciprocating cutter speed is given by:

$$N = \frac{12 \times V \times \cos H}{L \times \pi}$$

where: V = cutting speed in surface feet per minute
H = gear helix angle
L = length of stroke in inches
π = 3.14159...
N = number of strokes per minute

For spur gears $H = 0$ and $\cos H = 1$ and the equation becomes:

$$N = \frac{12 \times V}{L \times \pi}$$

The reciprocating cutter speed is not infinitely variable on conventional gear shapers. After calculating the theoretical value of N, an actual value must be selected in the range of available speeds which is nearest to the theoretical value.

Modern machines have the capability of changing spindle speeds between roughing and finishing cuts. Machines equipped with pole switching motors can double the number of strokes per minute for finishing, while CNC shapers have infinitely variable cutting speeds.

The maximum allowable cutting speed is determined by:

* Machinability of workpiece material;
* Shaper cutter material;
* Condition of machine and tooling;
* Desired toollife;
* Workpiece quality;
* Coolants.

Machinability rating and shaper cutter material are the two most important factors affecting speeds and feeds. Below is a table of recommended speeds and feeds for various materials machined with HSS cutters. Titanium nitride coating allows for 30 to 40% increase in cutting speed.

Table 3-1
Cutting Speeds for HSS Cutters

Material	Brinell Hardness	Cutting Speed	
		Rough feet/min	Finish feet/min
SAE 8615	161 - 205	92 - 115	184 - 230
SAE 8620	170 - 217	98 - 115	197 - 230
SAE 8630	179 - 229	98 - 115	197 - 230
SAE 1015	131 - 170	98 - 130	197 - 262
SAE 1045	179 - 229	98 - 115	197 - 230
SAE 1065	183 - 241	98 - 115	197 - 230
SAE 5140	174 - 229	92 - 105	184 - 210
SAE 4135	187 - 229	92 - 105	184 - 210
SAE 4130	187 - 229	98 - 115	197 - 230
Cast iron	220 - 240	66 - 98	131 - 197
Cast iron	193 - 220	66 - 82	131 - 164

Table 3-2
Generating Feeds for HSS Cutters

Material	Brinell Hardness	generating feed in inch/stroke				
		Diametral pitch				
		25 - 16	16 - 10	10 - 6	6 - 4	4 - 2
SAE 8615	161 - 205	0.008	0.010	0.012	0.016	0.020
SAE 8620	170 - 217	0.008	0.010	0.012	0.016	0.020
SAE 8630	179 - 229	0.008	0.010	0.012	0.016	0.020
SAE 1015	131 - 170	0.008	0.010	0.012	0.016	0.020
SAE 1045	179 - 229	0.008	0.008	0.010	0.014	0.020
SAE 1065	183 - 241	0.008	0.010	0.012	0.016	0.020
SAE 5140	174 - 229	0.008	0.008	0.010	0.012	0.016
SAE 4135	187 - 229	0.008	0.008	0.010	0.012	0.016
SAE 4130	187 - 229	0.008	0.010	0.012	0.016	0.020
Cast iron	220 - 240	0.006	0.007	0.008	0.012	0.016
Cast iron	193 - 220	0.008	0.010	0.012	0.016	0.020

Length of Stroke

The stroke length consists of:

* Gear face width,
* Approach, and
* Overrun.

The gear or spline face width is a value which is read directly from the part print. Approach and overrun are only required to ensure safe cutting conditions. As a rule of thumb, 15% should be added to the face width to account for approach and overrun distances (Figure 3-1).

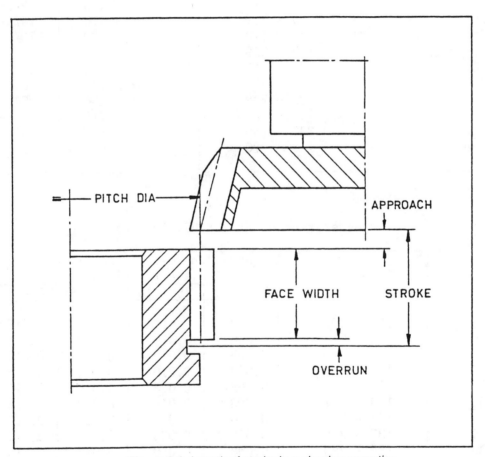

Figure 3-1. Length of stroke in a shaping operation.

Cycle Time Formulas

Gears and splines can be shaped in one, two, or three subsequent cuts, depending on the size of the tooth, with varying speeds and feeds for each cut. Different cycle time calculations are derived for one, two, or three cut cycles based on number of cuts taken and cutting parameters used during each cut.

Single Cut Cycle

The gear is cut in one revolution of the workpiece. The cycle consists of infeed and rotary feed and is expressed as:

$$T = \frac{D + C}{Fi \times N} + \frac{PCD \times \pi}{Fg \times N}$$

where:
- T = shaping time in minutes
- D = depth of cut in inches
- Fi = infeed in inches per stroke
- N = number of cutter strokes per minute
- PCD = pitch circle diameter of gear in inches
- π = 3.14159
- Fg = generating feedrate in inches per stroke
- C = cutter clearance

Two Cuts Cycle

This is the most frequently used shaping cycle, the toothform is finished in two revolutions of the workpiece, and it consists of a roughing and a finishing pass. Cutting speed is increased during the second pass. Feedrates can be increased or decreased based on the quality specifications (Figure 3-2).

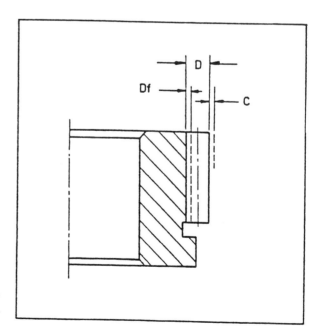

Figure 3-2. Depth of cut in a shaping cycle consisting of a roughing and finishing pass.

	infeed	rotary feed
1st revolution	$\dfrac{D - Df + C}{Fi \times Nr}$	$\dfrac{PCD \times \pi}{Fgr \times Nr}$
2nd revolution	$\dfrac{Df}{Fi \times Nf}$	$\dfrac{PCD \times \pi}{Fgf \times Nf}$

Figure 3-2. Continued.

$$T = \frac{D - Df + C}{Fi \times Nr} + \frac{PCD \times \pi}{Fgr \times Nr} + \frac{Df}{Fi \times Nf} + \frac{PCD \times \pi}{Fgf \times Nf}$$

where: Nr = strokes/minute for rough cut
Nf = strokes/minute for finish cut
Fgr = generating feedrate for rough cut
Fgf = generating feedrate for finish cut
Df = finishing stock

The cycle consists of: - feeding the cutter to distance Df of full depth at speed Nr and feed Fi - one rough generating cut at speed Nr and feed Fgr - feeding the cutter over distance Df, to full depth, at speed Nf and feed Fi - one finish generating cut at speed Nf and feed Fgf.

Three Cuts Cycle

Assuming that two roughing cuts are taken, each removing half of the tooth depth $D/2$ at speed Nr and generating feed Fgr, and one finishing cut to remove finishing stock Df at speed Nf and feed Fgf, the cycle is calculated as follows (Figure 3-3).

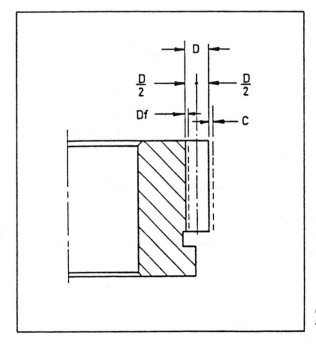

Figure 3-3. Depth of cut in a three cuts shaping cycle.

	infeed	rotary feed
1st revolution	$\dfrac{D/2 + C}{Fi \times Nr}$	$\dfrac{PCD \times \pi}{Fgr \times Nr}$
2nd revolution	$\dfrac{D/2 - Df}{Fi \times Nr}$	$\dfrac{PCD \times \pi}{Fgr \times Nr}$
3rd revolution	$\dfrac{Df}{Fi \times Nf}$	$\dfrac{PCD \times \pi}{Fgf \times Nf}$

Figure 3-3. Continued.

$$T = \frac{D/2 + C}{Fi \times Nr} + \frac{PCD \times \pi}{Fgr \times Nr} + \frac{D/2 - Df}{Fi \times Nr} + \frac{PCD \times \pi}{Fgr \times Nr} + \frac{Df}{Fi \times Nf} + \frac{PCD \times \pi}{Fgf \times Nf}$$

$$= \frac{D/2 + C + D/2 - Df}{Fi \times Nr} + \frac{2 \times PCD \times \pi}{Fgr \times Nr} + \frac{Df}{Fi \times Nf} + \frac{PCD \times \pi}{Fgf \times Nf}$$

$$T = \frac{D + C - Df}{Fi \times Nr} + \frac{2 \times PCD \times \pi}{Fgr \times Nr} + \frac{Df}{Fi \times Nf} + \frac{PCD \times \pi}{Fgf \times Nf}$$

Examples of Cycle Time Calculations

Example 1: Single Cut Shaping Cycle for External Involute Spline.

Part print data

diametral pitch	27.6658
pitch diameter	4.5543
major diameter	4.5905
minor diameter	4.4885
spline width	1.060
material	SAE 8620

Machine setting

Cutting speed	100 SFPM
Feedrate infeed	0.002 in./stroke
Generating feedrate	0.014 in./stroke
Cutter clearance	0.020 in.
Single cut cycle	

Cycle time calculation

Stroke length: $1.060 \times 1.15 = 1.22$ inch

Number of strokes per minute $= \dfrac{12 \times 100}{1.22 \times 3.14159} = 313$

300 strokes/minute is the nearest available speed on the machine.

$$\text{Toothdepth} = \frac{4.5905 - 4.4885}{2} = .051 \text{ inch}$$

$$\text{Cycle time} = \frac{.051 + .020}{.002 \times 300} + \frac{4.5543 \times 3.14159}{.014 \times 300}$$

$$= .118 + 3.406 = 3.524 \text{ min}$$

Example 2: Two Cuts Shaping Cycle for Internal Ringgear

Part print data

diametral pitch	5.08
pitch diameter	8.0708
toothdepth	0.445
gear face width	0.866
material	SAE 1040

Machine setting

Cutting speed roughing	100 SFPM
Cutting speed finishing	200 SFPM
Feedrate infeed	0.001 in./stroke
Generating feed roughing	0.020 in./stroke
Generating feed finishing	0.014 in./stroke
Cutter clearance	0.020 in.
Finishing stock	0.010 in.

Cycle time calculation

Stroke length: $.866 \times 1.15 = .995$ inch

$$\text{Strokes per minute roughing} = \frac{12 \times 100}{.995 \times 3.14159} = 383$$

355 is nearest speed on the machine. Finish shaping is performed at 710 strokes/minute.

$$\text{Cycle time} = \frac{.445 - .010 + .020}{.001 \times 355} + \frac{8.0708 \times 3.14159}{.020 \times 355} + \frac{.010}{.001 \times 710} + \frac{8.0708 \times 3.14159}{.014 \times 710}$$

$$1.281 + 3.571 + .014 + 2.551 = 7.417 \text{ minutes}$$

Field of Application

* Shaping is a rapid and economic gear cutting method suitable for high and low volume production, and capable of producing a wide variety of tooth forms like, spur and helical gears, splines, herringbone gears and gear segments.

Shaping is mainly used for internal gears and shoulder gears, which are gear configurations not suitable for hobbing. Sometimes, the use of the shaping process is dictated by design considerations. The hob runout beyond the active face width of the gear is not always permissible (Figure 3-4) The small undercut of approximately 1/8 in. (3.2 mm) required for shaping provides in many instances for a better design solution.

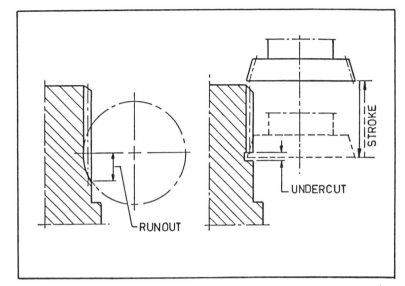

Figure 3-4. The shaping process is selected when cutter runout marks in an adjacent diameter are not permissible.

The shaping process is limited by its length of cut which is shorter than hobbing. Productivity is inherently less than hobbing, because no cutting occurs during the return stroke. Stacking of parts is not advantageous because there is no approach length in shaping.

The shaping process is however capable of producing more accurate gears than hobbing because shaper cutters can be ground to higher quality standards than hobs.

Taper Shaping

A tapered tooth form can be obtained by tilting the axis of the shaper cutter (Figure 3-5), or by tilting the axis of the workgear. Tapered splines are often used in synchronizer and coupler designs to prevent jumping out of gear. A tapered tooth form also compensates for heat treatment distortions and unequal shrinkage.

Combination shaping

Two gears or splines with the same number of teeth can be cut simultaneously with two shaper cutters mounted on the same arbor (Figure 3-6). The tooth forms may differ in pressure angle, or tooth thickness but must have the same number of teeth. This setup requires only half the stroke length of a single cutter operation. Machining time is, therefore, reduced by 50%.

Cutting of Two Gears with Same Diametral Pitch

This is a typical CNC shaping operation where two gears with same diametral pitch and same pressure angle are cut in sequence during the same cycle (Figure 3-7). The machining parameters are adjusted automatically for each gear by the CNC control so both gears are produced in optimum cutting conditions. It is essential for this application that one shaper cutter design suits both gears.

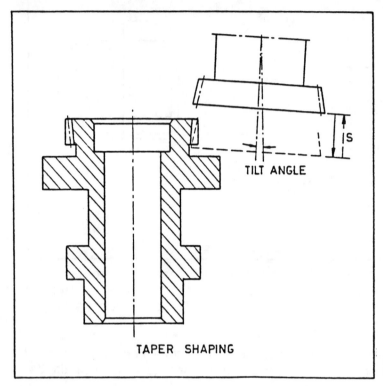

Figure 3-5. Shaping of tapered tooth form.

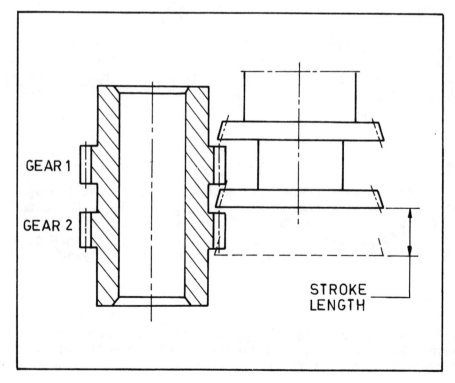

Figure 3-6. Combination shaping of two gears in one stroke of the spindle. Two cutters are mounted on an arbor with a spacer between the cutters.

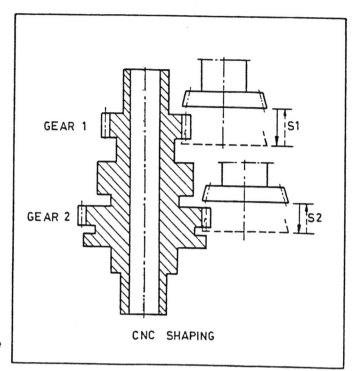

Figure 3-7. Consecutive shaping of two gears in one machine cycle, using the same cutter.

Cutting of Two Gears with Different Diametral Pitch

When the gears have different diametral pitches, the same cutter can obviously not be used. The solution is to mount two cutters on the same arbor as shown in Figure 3-8.

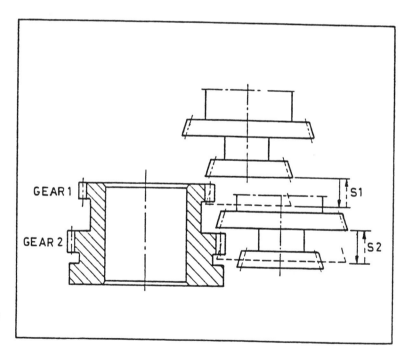

Figure 3-8. Consecutive cutting of two gears in one machine cycle, using two different shaper cutters, mounted on an arbor.

Cutting Two Gears with Up and Down Shaping

The clustergear shown in Figure 3-9 can be cut in one cycle by mounting two shaper cutters back to back on the same arbor. The upper gear is cut with downstroke, while the lower gear is cut with upstroke.

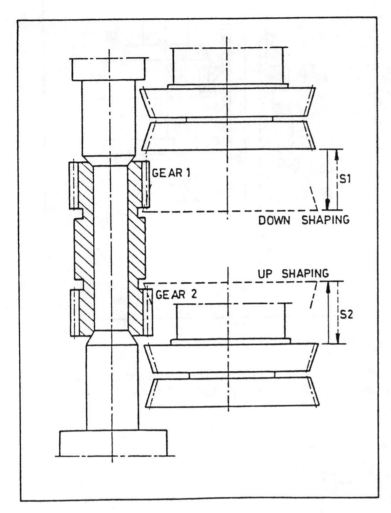

Figure 3-9.
Consecutive cutting of two gears, using back to back mounted shaper cutters. The direction of the cutting stroke is reversed for each gear.

Cutting of External and Internal Gears

An external gear and an internal gear can be cut in the same cycle by mounting internal and external shaper cutters on the same arbor (Figure 3-10). Machining parameters for each cut are programmed in the CNC control.

Cutting of External Recessed Tooth Forms with an Enveloping Cutter

The gear shown in Figure 3-11 has an external spline tooth which is recessed below the rim face of the gear. With an external shaper cutter, the cutter diameter would be limited to maximum 1.00 in. (25.4 mm) resulting in low generating feeds, multiple passes, long cycle times, and low cutter life. Tool life can be increased, and cycle time reduced, by using an enveloping shaper cutter with internal teeth. The spline teeth can be cut straight, or with backtaper as shown in Figure 3-11 on a tilting table mechanism. Accuracy of an enveloping cutter is inherently less because of the larger manufacturing tolerances required for the internal teeth.

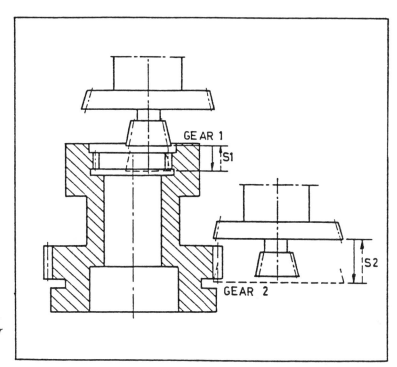

Figure 3-10. Consecutive cutting of an internal gear and an external gear in the same machine cycle.

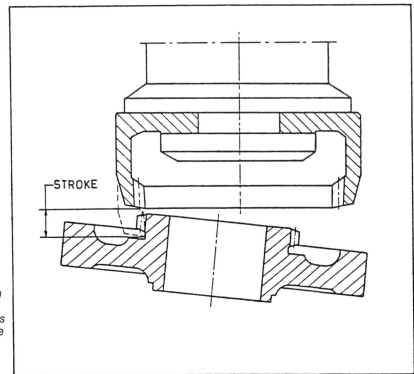

Figure 3-11. Shaping of a recessed tooth form with an enveloping cutter. The tooth form is tapered by tilting the axis of the workpiece relative to the axis of the shaper cutter.

4
Shaving
THE PROCESS

The shaving process can be compared to two gears which rotate in tight mesh while one gear traverses across the face of the other. The shaving cutter is driven by an electric motor and, in turn, drives the work gear which rotates freely between centers. The axial feed is accomplished by a traversing motion of the worktable on which the work gear is mounted. At the end of each stroke the direction of feed is changed and the cutter rotation is reversed.

The shaving cutter is basically a helical gear with a helix angle different than that of the gear to be shaved. Serrations in the faces of the teeth form the cutting edges.

During the shaving operation, the axes of work gear and cutter are crossed at an angle, which is the difference between the helix angle of cutter and gear. The crossed axes angle is the cause of the cutting action in shaving.

Infeed is accomplished by moving the centers of cutter and gear closer together. The tooth thickness is reduced by moving the center of the work gear upward, closer to the cutter. Infeed causes pressure between cutter and work gear, the cycle is therefore completed by two finishing strokes without vertical feed.

The four different methods of shaving are:

* Conventional shaving, when the traversing motion of the workpiece is parallel to its axis.
* Diagonal shaving, when the traversing motion of the workpiece is at an angle with its own axis.
* Underpass shaving, the workpiece moves at an angle of 90 degrees relative to its own axis.
* Plunge shaving, in this method there is no traverse movement between gear and cutter. The cutter is fed gradually to full depth by reducing the center distance between cutter and gear.

Shaving is a fast and universal method of gear finishing. It is applicable to external and internal gears. However, the machines used for external and internal shaving are usually not interchangeable.

Practically, the shaving process is limited by the size of the workpieces and the quality specifications. Since shaving is performed prior to heat treatment it is not applicable to high accuracy gears which are usually finished by grinding or honing.

Cycle Time Formula

The formula to calculate shaving cycle time is:

$$T = \frac{N \times L}{F}$$

where: T = cycle time in minutes
L = stroke length in inches
F = feedrate in inches per minute
N = number of strokes

Cutting Speed

The cutting speed is defined as the peripheral velocity of the shaving cutter.

$$V = \frac{Nc \times Dc \times \pi}{12}$$

where: V = cutting speed in feet per minute
Nc = cutter revolutions per minute
Dc = cutter diameter in inches
π = 3.14159

The number of rotations of the cutter is given by the formula:

$$Nc = \frac{12 \times V}{\pi \times Dc}$$

the number of rotations per minute of the workgear is:

$$Ng = \frac{Nc \times Zc}{Zg}$$

where: Ng = revolutions per minute of gear
Zc = number of teeth in cutter
Zg = number of teeth in gear

Ng can also be found by multiplying Nc with the ratio:

$$\frac{Dc}{Dg}$$

where: Dg = Gear pitch diameter

Cutter sizes are approximated based on following standard data (Table 4-1).

Table 4-1
Data for Approximating Cutter Sizes (1)

Diametral Pitch Range	Cutter Pitch Dia	Standard Width
20 - 64	2.500 - 3.499	.500
16 - 36	4.500 - 5.499	.625
7 - 16	6.500 - 7.499	.750
4 - 10	8.500 - 9.499	1.000
2 - 4	11.500 - 12.499	1.000

The optimum cutting speed for shaving operations can only be determined by practice. However, following starting values will normally give satisfactory results.

Fine pitch gears with diametral pitch > 8 V = 400 SFM
Medium pitch gears with diametral pitch < 8 V = 350 SFPM
Coarse pitch gears with diametral pitch < 3 V = 275 SFPM

When shaving small diameter pinions, the speed of the pinion should be limited to 800 rpm maximum.

Example:

Gear   diametral pitch: 5.5
       pitch diameter: 3.307
       number of teeth: 16

Cutter diameter   9.0

$$\text{Cutter rpm} = \frac{12 \times 350}{3.14159 \times 9.0} = 148$$

$$\text{Gear rpm} = \frac{148 \times 9.0}{3.307} = 403$$

Cutting Feed

There is no fixed rule to determine the feedrate for a particular shaving operation, because it depends on many factors such as material, surface finish, hardness, diagonal traverse angle, cutting oil, etc.

(1) Modern Methods of Gear Manufacture, National Broach & Machine Division, Lear Siegler, Inc., Fourth Edition, 1972.

A starting feed of 0.010 in. (0.25 mm) per revolution of the gear is used in most cases. Actual feedrates will, of course, vary depending on cutting conditions and machinability of the workgear material.

When shaving shoulder gears with a low crossed axis angle, cutting action is limited and feedrate should be reduced accordingly. Feeds and speeds must also be reduced when shaving material with low machinability rating or pre-heat treated steel.

On a conventional shaving machine speeds and feeds are set by means of change gears.

The shaving feedrate is given by the formula:

$$F = Fn \times Ng$$

where: F = feedrate in inch per minute
Fn = feed per revolution of workpiece
Ng = workpiece rpm

In the previous example Ng = 403 rpm.
Assuming Fn = 0.010 in. (0.25 mm)/revolution the feed is
F = .010 X 403 = 4.03 in. (102 mm) per minute.

Stroke Length

The stroke length varies with the shaving method.

For conventional shaving the stroke length equals the face width of the gear (Figure 4-1).

$$L = FW \text{ (gear face width)}$$

A small amount of overtravel clearance must be added to the stroke length to ensure full clean up of the gear face, 0.060 in. (1.5 mm) per side, so that the actual stroke length is L +0.120 in. (3.0 mm).

For diagonal shaving the stroke length is shown in Figure 4-2.

$$L = AC \times \cos(DT) - BC \times \cos(DT + CA)$$

$$L = FW \times \cos(DT) - FC \times \cos(DT + CA)$$

where: L = stroke length in inches
FW = gear face width in inches
CA = crossed axis angle
DT = diagonal traverse angle
FC = cutter face width in inches

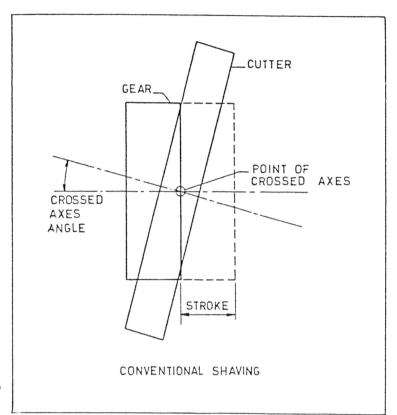

Figure 4-1. Stroke length in conventional shaving.

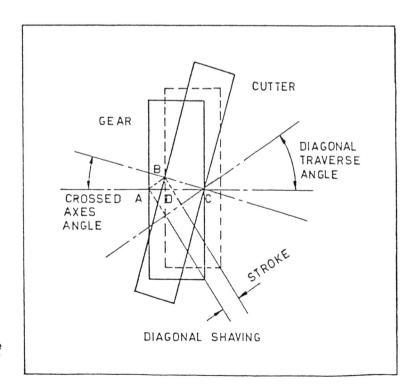

Figure 4-2. Stroke length in diagonal shaving.

With a given cutter width FC the maximum traverse angle DT is calculated as follows:

$$\tan(DT) = \frac{BD}{AD}$$

$$\tan(DT) = \frac{FC \times \sin(CA)}{FW - FC \times \cos(CA)}$$

While the minimum cutter width required for diagonal shaving can be found with:

$$\min FC = \frac{FW}{\dfrac{\sin(CA)}{\tan(DT)} + \cos(CA)}$$

A narrow faced cutter and wide gear will limit the diagonal traverse angle to a small value. The traverse angle can be enlarged by increasing the width of the shaving cutter. A shaving cutter slightly wider than the gear permits shaving with 90 degrees traverse angle.

A crossed axis angle between 10 and 15 degrees suits practically all shaving setups. Shoulder gears may require a crossed axis angle below 10 degrees with 5 degrees as minimum.

A diagonal traverse angle between 20 and 40 degrees will give the best results with respect to tool life and cycle time.

Practically all gears with face width less that two inches (50.4 mm) can be shaved conventionally or diagonally. Shoulder gears may only be suitable for diagonal shaving depending on the clearance between gears.

As can be seen from the mathematical expression, diagonal shaving is the fastest method because of the shorter stroke length. Wide gears, with a face width over 2.0 in. (51 mm), are shaved conventionally as diagonal shaving would require a wide and expensive shaving cutter. Cluster gears must be separated by a minimum gap to allow sufficient travel to finish the whole gear width (Figure 4-3). Usually, a 3/8 in. (9.5 mm) gap is sufficient to clear the cutter. The diagonal shaving angle can be increased to 80 to 85 degrees for close shoulder gears.

To estimate the machine cycle time an effective feedrate is considered which is based on the speed at which the point of crossed axes moves across the face of the gear.

$$\text{Effective feed rate per gear revolution} = \frac{0.040}{\dfrac{\sin(DT) + \cos(DT)}{\tan(CA)}} \qquad (2)$$

The cycle time is found by dividing the stroke length by the effective feed.

$$T = \frac{N \times L}{F_E}$$

(2) Modern Methods of Gear Manufacture, National Broach & Machine Division, Lear Siegler, Inc., Fourth Edition, 1972.

where: T = cycle time in minutes
N = number of strokes
L = stroke length in inches
F_E = effective feed in inch/min
$\left(\begin{array}{l} \text{effective feed in inch/rev of gear multiplied} \\ \text{with the gear rpm.} \end{array} \right)$

Underpass shaving is a special case of diagonal shaving with a diagonal traverse angle of 90 degrees (Figure 4-4). Underpass shaving requires a shaving cutter which is slightly larger than the face width of the gear.

$$DT = 90 \text{ degrees and } \sin(90) = 1$$

$$L = FW \times \cos(90) - FC \times \cos(CA + 90)$$

$$= 0 - FC [\cos(CA) \times \cos(90) - \sin(CA) \times \sin(90)]$$

$$= -FC [0 - \sin(CA)]$$

$$L = FC \times \sin(CA)$$

Just as for diagonal shaving, an effective feedrate is considered along the contact point of crossed axes to estimate cycle time.

In the effective feedrate formula, $DT = 90$ and $\cos(DT) = 0$, $\sin(DT) = 1$

$$\text{Effective feedrate} = 0.040 \times \tan(CA)$$

Number of Strokes

The shaving cycle consists of a number of roughing and finishing strokes. The finishing strokes have no vertical upfeed and, in general, two finishing strokes are taken. Roughing strokes are performed with vertical upfeed at the end of each stroke. The amount of upfeed is determined by the upfeed cam in the machine. The number of roughing strokes depends on the shaving stock allowance.

The shaving stock per tooth flank is dependent upon the size of the gear tooth as shown in Table 4-2.

Table 4-2
Shaving Stock Per Tooth Flank and Size of the Gear Tooth

Diametral Pitch	Shaving stock per flank
17	0.0007
12.7	0.0008
10	0.0009
8.5	0.0010
7.2	0.0011
6.5	0.0012
5.6	0.0013
5	0.0014
4.6	0.0015
4.2	0.0016

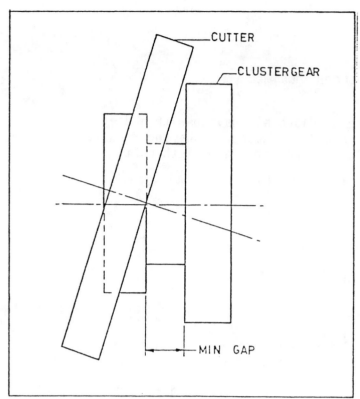

Figure 4-3. Minimum gap condition to shave shoulder gears.

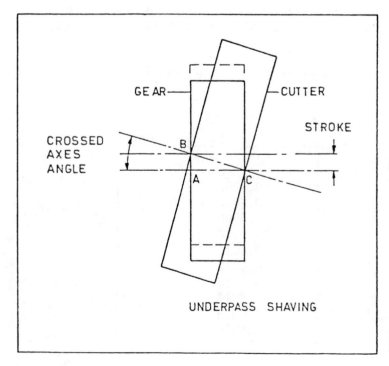

Figure 4-4. Stroke length in underpass shaving.

Usually the shaving cycle consist of six passes, four roughing passes with vertical upfeed and two finishing passes without upfeed. The stock removal per pass is therefore one fourth of the total stock per tooth flank. Progressively diminishing upfeed increments also are possible by cam adjustment.

As an example, for diametral pitch 8.5 the shaving stock per flank is 0.0010 in. (0.25 mm), and each of the four consecutive passes with upfeed will remove 0.00025 if the stock removal is equally divided among the four passes.

EXAMPLES OF CYCLE TIME CALCULATIONS

Example 1:

Gear data	diametral pitch	8
	pitch diameter	3.750
	number of teeth	30
	face width	0.880
Cutter data	diameter	9.0
	number of teeth	72
	face width	0.75
Cutting speed		400 SFPM
Feed		0.010 IPR
Conventional shaving		

Calculations

$$\text{Cutter rpm} = \frac{12 \times 400}{3.14159 \times 9} = 169$$

$$\text{Gear rpm} = \frac{169 \times 72}{30} = 405$$

Feedrate = .010 × 405 = 4.05 inch/minute

With .12 overtravel both sides of the gear face the stroke length = 0.88 + .24 = 1.12 inch

$$\text{Cycle time} = \frac{1.12 \times 6}{4.05} = 1.659 \text{ minutes}$$

Example 2

Same gear as above but with diagonal shaving process. Assume a crossed axis angle of 12 degrees and diagonal traverse angle of 40 degrees. The cutter width is 0.750 in. (19.05 mm).

$$\text{Effective feedrate} = \frac{0.040}{\dfrac{\sin(40) + \cos(40)}{\tan(12)}}$$

$$= \frac{0.040}{\dfrac{.64279 + .76604}{.21256}}$$

$= .0105$ inch per revolution of workgear

Feedrate in inches per min = 0.0105×405
= 4.252 inch per minute

Stroke length = $.88 \times \cos(40) - .750 \times \cos(40 + 12)$

$= .88 \times .76604 - .750 \times .61566$

$= .67411 - .46174$

$= .212$ inch

With .120 overtravel per side, the stroke is .212 + .24 = .452 inch.

$$\text{Cycle time} = \frac{0.452 \times 6}{4.252} = 0.638 \text{ minutes}$$

Example 3

The same gear cut in underpass shaving with a 1.00 in. (25.4 mm) wide shaving cutter will give following cycle time.

Stroke length = 1.0 × sin (12)

$\qquad$ = .208 inch

Because the shaving cutter is wider than the gear in underpass shaving, overtravel is reduced by approximately 30%.

.208 + .160 = 368 inch.

The effective feed is = .040 × tan (12)

$\qquad$ = .040 × .21256

$\qquad$ = .0085 inch per revolution

The feed in ipm is .0085 × 405 = 3.443

$$\text{Cycle time} = \frac{.368 \times 6}{3.443}$$

$\qquad$ = .641 min

Note: The cycle time calculations in the three examples are based on theoretical starting values for feeds and speeds and estimated values for crossed axes and diagonal traverse angles. Practically, when the cutting parameters are optimized for each method, the cycle time for diagonal shaving will be approximately half the conventional shaving time, while underpass shaving will show a further reduction of about 20% relative to diagonal shaving.

Field of Application

Shaving is a gear finishing method which is executed prior to heat treatment and after hobbing or shaping. The primary purpose of shaving is to improve the gear profile (involute) and surface finish. It can also correct errors in index, helix, and eccentricity. However the purpose of shaving is not to remedy carelessness in hobbing or shaping.

The shaving process is extensively used to modify the tooth profile to improve the durability of heavily loaded gears. Tooth form modifications can be made in the longitudinal direction or across the involute profile.

Crowning and Taper

Longitudinal crowning (Figure 4-5) is obtained by rocking the worktable during the traverse, so the ends of the gear teeth move closer to the cutter. The result is a tooth form which is slightly thinner at the ends than in the middle of the tooth. This will ensure the load is transmitted in the middle of the gear teeth and that end loading problems are avoided when shafts are misaligned. In a similar way, the machine table can be set at an angle to shave a tapered tooth form.

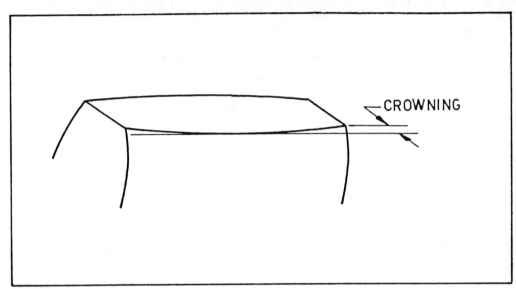

Figure 4-5. Longitudinal crowning obtained by shaving.

Involute Modifications

Tip and root relief (Figure 4-6) are modifications to the involute profile made with a specially ground cutter which removes slightly more material near the tip and the root of the gear tooth. Gear noise problems can often be resolved with tip and root relief.

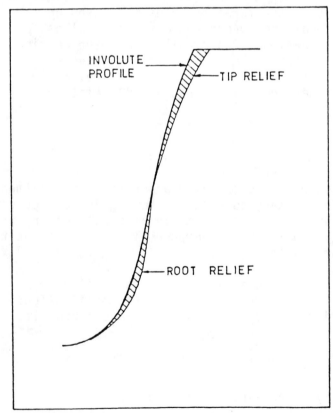

Figure 4-6. Involute corrections obtained by modifying the profile of the shaving cutter teeth.

Part II

Process Planning

Whether produced in large or small quantities, in a continuous flow line or in batches, the process planning sequence of gears can be broken down in four groups of operations.

* Blanking.
* Gear cutting.
* Heat treatment.
* Grinding.

"Blanking" is a term that refers to the initial machining operations that produce a semifinished part ready for gear cutting, starting from a piece of rough material. Turning on chuckers or lathes, face and centering of shafts, milling, and sometimes grinding, fall into this category of operations. Good quality blanks are essential in precision gear manufacturing

Machining operations can be divided into green-end and hard-end operations. Green-end designates all operations executed prior to heat treatment, when the material is still soft and easily machinable. Hard-end refers to operations after heat treatment, performed on parts which have acquired high hardness and strength.

In medium and high-volume production systems, the line flow concept is adopted for green-end and hard-end machining operations. Because of the special nature of the operation and dedicated support facilities, heat treatment is usually handled in bulk in a central area.

Heat treatment gives the material the strength and durability to withstand high loads and wear. The increased strength is, however, coupled to a reduction in dimensional and geometrical accuracy. The structural transformations which occur during hardening and the quenching strains are the cause of a general quality deterioration which must be anticipated and corrected in the process planning sequence.

External and internal bearing diameters, critical length dimensions, and fine surface finishes are obtained by precision grinding operations after heat treatment. Cylindrical grinders, angle head grinders, internal grinders, and surface grinders are commonly used machines in hard-end operations.

Hobbing, shaping, and shaving machines are the most frequently used machines in the gear industry. They produce gears for automotive, truck, agricultural, and construction equipment. A variety of other processes are used in industrial gear production as well. Broaching, rolling, grinding, milling, and skiving are processes which find cost-effective application based on quality specifications, production volumes, and economic considerations.

5
Gear Manufacturing Methods

PROCESS SELECTION

The process and type of equipment for a specific gear machining operation is selected by analysis of following workpiece characteristics.

* Part configuration,
* Gear quality specifications,
* Production quantity.

Part Configuration

The selection of the gear cutting method starts with an analysis of the drawing and the basic part configuration. The shape of the part is an important criterion because some methods of gear manufacturing are only applicable for parts with a specific configuration. For instance, hobbing and broaching require unobstructed axial travel of the cutting tool and, as a result, are not suitable for shoulder gears. Another example is gear rolling which, so far, has only been developed for external gears.

For the purpose of selecting the appropriate manufacturing method, part configurations can be classified according to the four types shown in Figure 5-1. Although the gears are shown as integral designs, the same classification applies for gears which are only an element of a cluster gear or a shaft. The two essential criteria are:

1. Whether the gear is internal or external, and
2. Whether the design is open ended or not. In an open ended design the gear cutting tool travel is not obstructed by an adjacent shoulder.

In Figure 5-1:

* Type A represents external gear, open ended,
* Type B represents external gear with adjacent shoulder,
* Type C represents internal gear, open ended,
* Type D represents internal gear with adjacent shoulder.

The field of application for each method is summarized in the Table 5-1.

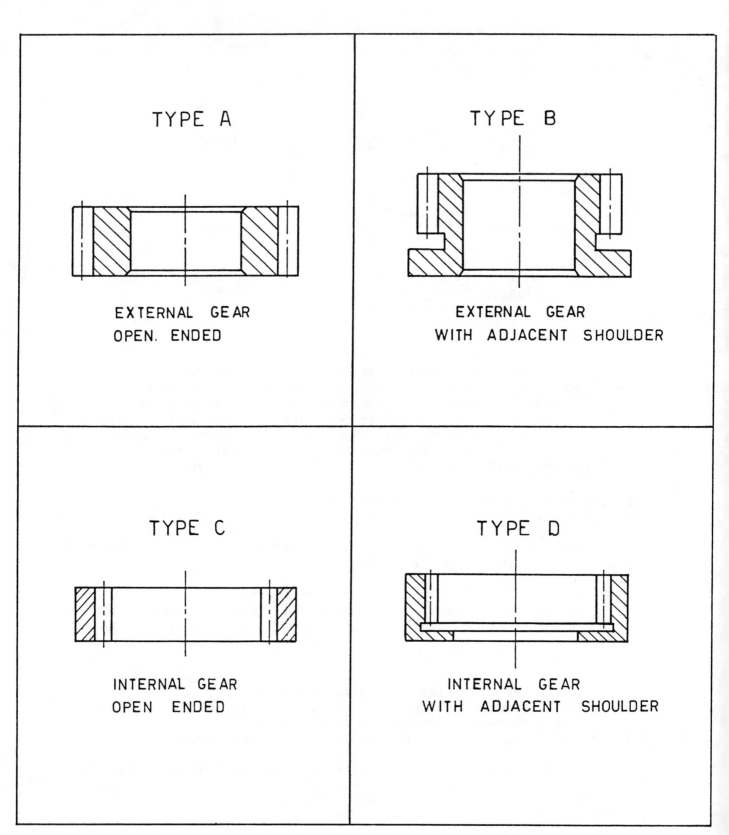

Figure 5-1. Classification of cylindrical gears based on part configuration.

Table 5-1
Field of Application

Method	Type A	Type B	Type C	Type D
Hobbing	yes	no	no	no
Shaping	yes	yes	yes	yes
Shaving	yes	yes	yes	yes
Milling	yes	no	no	no
Rolling	yes	yes	no	no
Internal broaching	no	no	yes	no
External broaching	yes	no	no	no
Grinding	yes	no	yes	no
Honing	yes	yes	yes	yes

Quality Specifications

In the United States, gear accuracy is expressed in AGMA (American Gear Manufacturers Association) quality classes which range from 3 to 15, with the highest number indicating the most accurate gear tolerances. German DIN standard gear quality classes have a correspondence with AGMA classes but run in reverse order from 11 to 3, with the highest number indicating the lowest level of accuracy. In both systems, the quality class specifies tolerances for involute, lead, tooth to tooth, and total runout for gears of a specific diametral pitch and within a range of pitch diameters.

Achievable tolerances are directly related to the manufacturing methods employed. A well-maintained machine with precision tooling, good blank preparation, and proper methods will produce more accurate gears than equipment operated with inadequate care.

The process capability of each method covers a range of quality classes and achievable accuracy will be consistent with the maintenance condition of equipment and tooling. However, there is a trade off between achievable quality and cost. A machine producing at the limit of its quality capability will run an unacceptable high cost for inspection, reject, repair, and scrap.

Good common sense in manufacturing implies that machines are utilized in their normal range of quality capability. Even if with the proper care an AGMA class 12 gear can be shaved within tolerances, it would be risky to assume that a high-volume shaving operation will deliver consistent quality class 12 gears. Since class 12 is borderline for shaving, another finishing method is desirable for trouble-free production.

Keep in mind also that the obtainable quality is specified for non-heat-treated gears. It is generally accepted that the quality deterioration caused by heat treatment is equivalent to two gear classes. A class 11 gear will fall in class 9 range after hardening.

Quality levels achievable with specific processes are summarized in the Table 5-2.

Table 5-2
Quality Levels Achievable with Specific Processes

```
AGMA        -   -  15  14  13  12  11  10   9   8   7   6   5   4   3
DIN         1   2   2   3   4   5   6   7   8   9  10  11  12   -   -
-------------------------------------------------------------------------
Hobbing                                 ZZZXXXXXXXXXXXX
Shaping                                 ZZZXXXXXXXXXXXXXX
Shaving                             ZZZXXXXXXXXXXXXXXX
Broaching                               ZZZXXXXXXXXXXXXXX
Rolling                                         ZZZXXXXXXXXXXXXXX
Grinding                XXXXXXXXXXXXXX
Honing                          ZZZZZXXXXXXXXX
Special         XXXXXXXXXXX
```

XXX: Average range
ZZZ: Less frequent application

High-precision gears require finish machining after heat treatment. The most common methods of gear hard finishing are form grinding, generating grinding, carbide hobbing, and honing.

Extremely high accuracy (DIN 1-2) can only be obtained with special processes in a controlled environment with temperature control, constant humidity, clean air, etc.

Production Volume

A production machine is conceptually designed to perform a specific task in a low or high-volume production environment. Machine and tooling can be flexible to allow for quick change over in a batch production line, or dedicated to one part and one operation in a continuous flow process. Various types of machines are offered on the market to satisfy the needs of low, medium, and high-volume manufacturers.

Gear machine concepts can be divided in three categories:

1. Generating machines,
2. Full form cutting machines working on one tooth at the time, and
3. Full form cutting machines working on all teeth simultaneously.

The first category of machines work according to the generating principle and use basic rack tooling or involute form tooling. This first category generates a gear tooth profile by a series of closely spaced cuts. Hobbing, shaping, generating grinding, and skiving belong to this category. Because of their versatility, these machines are universally applied in low and high-volume manufacturing.

The second category of machines use full form tools to cut one tooth at the time, resulting in low output process with long cycle times. These machines

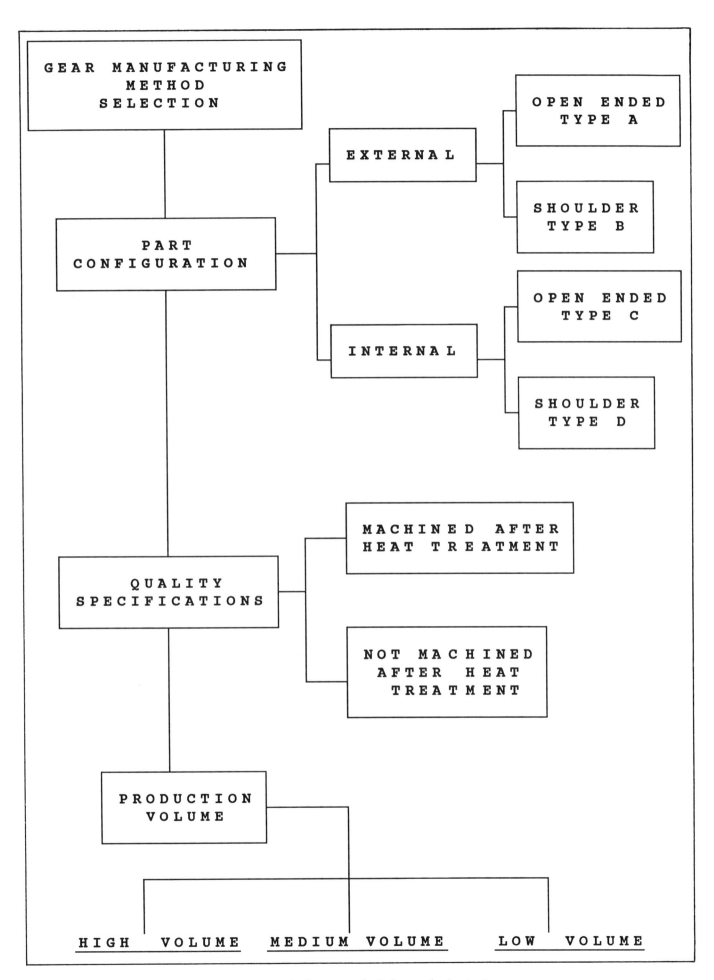

Figure 5-2. Gear manufacturing method selection.

are suitable for low and medium-volume batch runs. Milling and form grinding fall into this second category.

The third category of machines use specially designed full form tools to cut the whole contour of the gear simultaneously. The tools are designed for one specific gear. Because of their specialized nature, they are very expensive to manufacture and to maintain. The high capital investment in facilities and tooling is justified for high-volume, but is cost prohibitive for small production runs. Broaches and shear speed machines are examples of this category of machines.

Broaching machines for internal gears, and pot broaches for external gears are capable of cutting a gear to finished size in a single pass, in a cycle time less than 10 seconds.

OTHER GEAR PROCESSES

Gear Rolling

Unlike conventional gear cutting methods which remove metal, rolling is a chipless machining method which redistributes the surface metal along the tooth profile through a controlled cold forming process by applying localized pressure with a die. The rolling die is a precision ground external gear which meshes with the workpiece.

The surface layer of rolled material has a very smooth finish and superior grain structure properties which increases strength and durability.

Rolling cylindrical external gears has only been successfully applied as a finishing operation after the gear teeth have been rough cut with another method. The small pressure angles used in gear design and the height of the gear teeth, especially with coarse pitches, pose insurmountable problems when rolling from a solid blank.

As a finishing operation for fine pitch gears, rolling is extremely fast and ideally suited for high-volume production. Depending upon gear tooth characteristics and workpiece configuration, roll finishing can be completed in less than 10 seconds, excluding handling time.

The quality of finish rolled gears is influenced by the consistency of material hardness and ductility. It is also necessary to control the material allowance for rolling between very close limits to produce uniform parts. Good quality results have been obtained with finish rolling of planet gears for automotive transmissions.

Form Grinding

In form grinding, a grinding wheel diamond dressed to the proper profile, travels through one tooth space at the time and grinds two adjacent tooth flanks. After each pass of the grinding wheel, the work is rotated to the next tooth space by an indexing mechanism with a precision index plate (Figure 5-3).

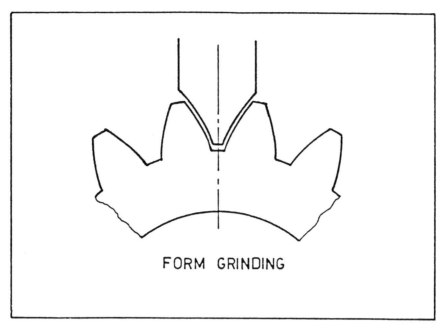

Figure 5-3. Finish grinding of one tooth space at a time with a form wheel.

Stock removal is controlled by moving the centers of grinding wheel and gear closer together. The work performs as many rotations as necessary to reach final size.

Form grinding produces high-quality precision gears. Lead and involute tolerances and surface finish are held within narrow specifications. Ovality and hardening distortions are completely removed. Index errors can be corrected within limits set by the accuracy of the indexing mechanism. Inevitably, the inherent manufacturing errors of the index plate will be repeated in the workpiece.

Form grinding is a flexible process adaptable to a wide range of parts. Involute modifications can be made by modifying the profile dressing template. Longitudinal crowning of the gear teeth is also possible with templates which control the distance between grinding wheel and workpiece during the longitudinal travel.

Internal gears are ground on machines equipped with special grinding spindles that reach inside the gear with a small diameter grinding wheel. The stock removal rate is as a result very low and internal gear grinding cycle times are extremely long.

Generating Grinding

Several methods of generating grinding have been developed based on basic rack grinding wheel profiles. The most productive method is the one using a worm shaped grinding wheel developed by Reishauer (Figure 5-4). The process is similar to hobbing except for the large diameter grinding wheel and the multiple passes taken across the face of the gear. Grinding occurs during both the up and down stroke of the wheel. Infeed is obtained by moving the centers of work and wheel closer together in small increments after each pass of the wheel.

Grinding with a worm wheel produces high-quality gears with accurate profile and index tolerances, and smooth surface finish.

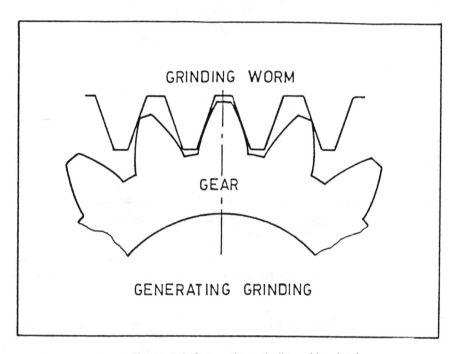

Figure 5-4. Generating grinding with a basic profile worm wheel.

Other methods of generating grinding developed by Maag and Hofler use a single or two flat grinding wheels to generate one tooth profile at the time. These machines are slower than the Reischauer gear grinders and are usually employed for large gears.

Just as with hobbing, generating grinding is limited by the gear configuration and is not applicable for shoulder gears. The limitation is even more stringent than for hobbing because grinding wheel diameters are much larger than hob diameters.

Tip and root relief are accomplished by changing the profile of the grinding worm. Reducing the distance between workpiece and grinding wheel produces longitudinal modifications of the gear teeth. To make crowned teeth the reduction in center-distance is controlled by templates mounted on the work slide.

Two start grinding worms can be utilized to increase productivity. Fine pitch gears, with tooth depth less than 0.032 in. (0.8 mm), can be ground from the solid.

Gear grinding is a time-consuming and costly operation. Cycle times may range from 5 to 20 minutes for truck and tractor transmission gears, and up to several hours per piece for larger gears. Besides high cost, an other concern regarding grinding is the danger of grinding cracks and surface burning.

Gear Honing

Gear honing is a hard finishing process effectively applied for heat-treated spur or helical gears to remove nicks and burrs, and to make minor improvements in involute profile and helix angle.

The main advantage of honing is an improvement in the sound quality of shaved and hardened gears by the removal of nicks, burrs, and other surface imperfections. Wear life is prolonged by the improvement in surface finish, and load carrying capacity is increased by higher surface contact between mating gears. To a limited extent the honing process can be used to salvage defective gears with excess runout or out-of-tolerance lead and involute conditions.

The working principle of the gear honer has a lot of similarity with the shaving process. The honing tool is a helical gear made of abrasive material which rotates with crossed axes, and in controlled mesh relationship with the workpiece. The honing tool traverses back and forth across the face of the gear, and direction of rotation is reversed at the end of each stroke. The workpiece rotates freely between centers, and cutting action is maintained by applying pressure between workpiece and honing gear.

The stock removal in honing is very small, 0.0002 to 0.0005 in. (0.005 to 0.012 mm) per tooth flank. This explains why only minor quality improvements can be attained with gear honing. Honing however has a beneficial effect on gear noise. A short honing cycle already creates a considerable improvement in noise level. Different types of honing machines are available for internal and external gears.

Gear Milling

This process is an adaptation of the conventional milling process to make gears on a universal milling machine equipped with an indexing head. A form milling cutter is used to cut one tooth space at the time in one or two passes. The gear is indexed to the next tooth space after every pass of the cutter until all teeth have been finished. Very large gears with diametral pitch 2 to 1 are form milled on indexing gear hobbing machines. The universal milling machine system with index head is mainly used in toolroom applications.

Internal Gear Skiving

Skiving is a gear generating method used to cut large internal ring gears on a hobbing machine equipped with a special skiving head. The skiving tool is in principle, a shaper cutter that rotates in continuous mesh with the work gear. The tool is fed axially down and cutting action is obtained by the generating movement of cutter and workpiece under crossed axes.

Skiving is a versatile process which offers the advantage of flexibility in low-volume production. A skived internal gear can be finish shaved on the same machine after changing cutters. Universal hobbing machines can be equipped with a skiving head interchangeable with a normal hobbing head. When skiving utilization is low the machine can still be operated as a conventional hobber for external gears.

Skiving machines are built for ring gears with a minimum internal diameter of 8 in. (203 mm) and maximum face width of 4 in. (102 mm). The capacity of large machines goes up to 100 in. (2540 mm) maximum internal diameter and 8 in. (203 mm) maximum face width.

Skiving with high-speed steel cutters, as a roughing operation, is faster than conventional gear shaping. Coarse pitch gears can be roughed and finished by the skiving process, better quality results are, however, obtained with other finishing methods like shaping or shaving. Skiving has been successfully applied in the industry to cut planetary ring gears for final drives of tractors and construction machinery.

Internal Gear Broaching

Broaching is an efficient, accurate method of producing internal ring gears which finds an economic application in high-volume manufacturing.

In the past, internal gear broaching was performed as a roughing operation followed by finish shaping or shaving. With full form finishing broaches, gears can now be produced to final size in a single pass of the broaching machine without any subsequent finishing operation.

Conventional broaches for internal gears are nibbling type broaches. The nibbling broach has teeth with progressively increasing height which generate the involute profile by a series of progressively deeper cuts. The involute profile generated is the result of numerous small and closely spaced cuts.

With this method, an accurate gear profile can only be obtained by maintaining alignment and perpendicularity of the machine, and by quality regrinds of the nibbling broach. If machine misalignment or non uniform tooth dulling causes broach off-center drift, the involute profile will be deformed.

The full form broach consists of a roughing section and finishing section. The roughing section is a conventional nibbling type broach, followed by a full form shaving shell that removes material along the full contour of the tooth. The full form broach cuts an accurate tooth profile with smooth surface finish (Figure 5-5).

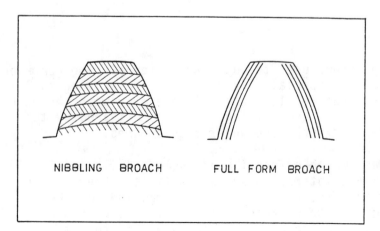

Figure 5-5. Difference in cutting action between nibbling broach (used for roughing), and full form broach, (used for finishing).

The main advantage of broaching is the high output and productivity combined with uniform and consistent quality. The output of a broaching machine is determined by the cutting speed, the broach length, and the speed of the load and unload system. Cutting speed is in the range of 10 to 30 surface feet per minute with high-speed steel broaches.

Broach cycles are so short that part handling automation has a major impact on the pieces per hour as shown in Table 5-3.

Table 5-3
Broach Cycle Examples

	Manual load	Automatic load
Broach and return stroke	15 sec	15 sec
Load and Unload	15 sec	2 sec
Cycle time	30 sec	17 sec
Pieces per hour @ 100%	120	211

The disadvantages of broaching are the high cost of the tooling, and each part requiring a special broach. Obviously, the process is economically not feasible for low-volume batch production.

Development of the correct tooth profile in conjunction with heat-treatment is also more elaborate than with basic rack tooling. All corrections for heat-treatment distortions must be ground in the profile of the broach.

Internal gear broaching has been successfully applied for spur gears as well as helical gears.

External Gear Broaching

External gear broaching, or potbroaching, is the fastest method available today for production of high-volume external gears. The potbroach forms the complete gear in a single pass of the part through the machine. Production rates of 250 per hour are not uncommon.

In modern potbroaching machines, the broaching tool is held stationary, upside down, and the part is either pulled or pushed up through the broach. With this setup chips fall freely away from the broach and coolant flow to the cutting edges is never obstructed by the part.

The pullup system is used when part diameter and length of the broach are such that a push rod would not have sufficient strength to support the workpiece rigidly. A pullup rod remains straight under load and provides better support than an equal diameter push rod.

Potbroaching of external gears is applied on a limited scale in precision gear manufacturing because of the difficulties encountered in making the broaching tool within close tolerances.

The potbroaching tool is composed of separate high-speed steel roughing, and finishing sections which are ground separately and inserted in a fixture.

The manufacturing tolerances of broach inserts and fixtures are critical for spacing and runout errors of the gear.

Another disadvantage is that broaching can only produce parallel tooth forms, longitudinal crowning of the teeth, which is desirable in many applications, cannot be attained.

The length of the broach can pose additional limitations to the broaching process. The broach length is determined essentially by the amount of stock it has to remove (tooth depth of the gear), the length of the cut (gear width), and by the chip thickness for the material to be processed. Coarse pitch gears, with large tooth width require such a long broach as to make the process no longer feasible.

Finally, internal or external broaching is restricted by the linear motion of the part relative to the tool. A nearby shoulder would obstruct the linear travel, making broaching only suitable for open-ended tooth forms.

Carbide Finish Hobbing

Carbide finish hobbing, also referred to as skiving, is a technique for finishing gears after heat-treatment and is, as such, an alternative to gear grinding.

In essence, skive hobbing differs only from conventional hobbing with regard to the cutting tool material and geometry. A solid carbide hob with negative rake angle is used for the operation. The achievable quality is comparable with normal grinding quality, surface finish is, however, less because of the typical hob tool marks.

Cutting speeds up to 450 surface feet may be used depending on tooth size and hardness, feedrates range from 0.060 to 0.100 in. (1.5 to 2.5 mm) per revolution of work piece. Stock removal is usually around 0.008 to 0.010 in. (0.2 to 0.254 mm) per flank.

Compared to grinding, carbide finish hobbing has the advantage of shorter cycle times. The cost of the solid carbide hob, and the regrinding cost are, however, very high and limit the application to special cases. To improve surface finish and noise performance, carbide hobbed gears can be finish honed.

6
Spline Manufacturing Processes

PROCESS SELECTION

The processing principles for gears are also applicable to spline manufacturing. With the appropriate cutting tools, gear machinery can also be used to cut spline teeth. However, a number of methods and machines have been specially developed for efficient manufacturing of splines.

The most suitable spline manufacturing process is selected based on three criteria:

* the basic part configuration,
* the tooth geometry, and
* the required production rate.

Part Configuration

Similar to gears, spline configurations are grouped in four basic types (Figure 6-1):

* Type A : external spline, open-ended
* Type B : external spline with adjacent shoulder
* Type C : internal spline, open-ended
* Type D : internal spline with adjacent shoulder

The applicability of each process based on the shape of the part is summarized in Table 6-1.

Tooth Form

The form of the spline tooth affects the design of the cutting tools in conjunction with the production process. According to tooth form, splines can be categorized as either involute splines, straight-sided splines, or serrations (Figure 6-2).

Among involute splines a further distinction is made for 45 degree pressure angle splines because of the special application of the spline rolling process.

Splines are used in driveline design to provide a solid coupling between rotating parts. Mating splines are assembled with a sliding fit which guarantees interference free assembly with minimum play.

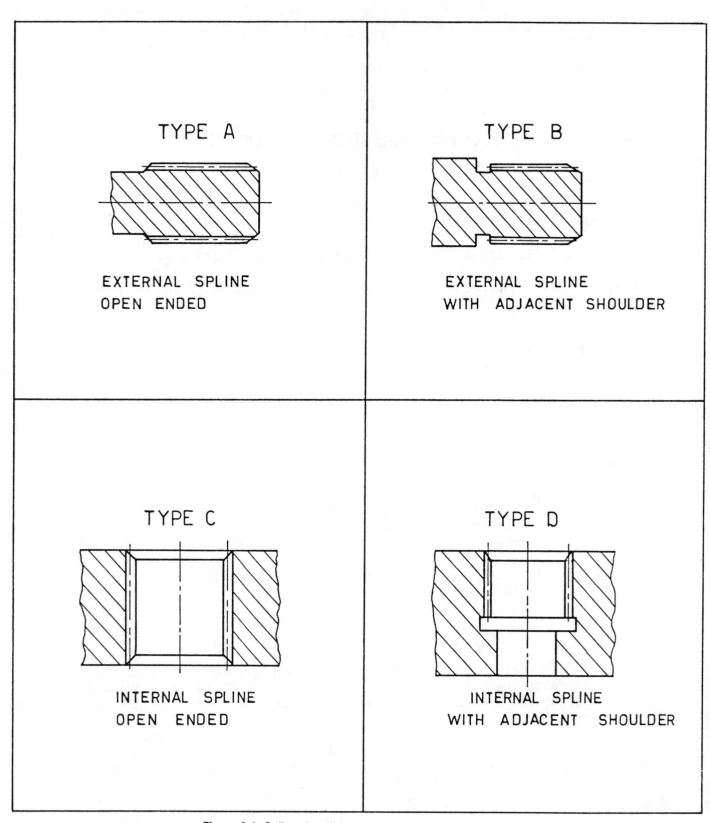

Figure 6-1. Spline classification based on part configuration.

Table 6-1
Process Applicability Based on Part Shape

Method	Type A	Type B	Type C	Type D
Hobbing	yes	no	no	no
Shaping	yes	yes	yes	yes
Shaving	yes	yes	no	no
Milling	yes	no	no	no
Rolling	yes	yes	no	no
Internal Broaching	no	no	yes	no
External Broaching	yes	no	no	no
Progressive Blind Broaching	no	yes	no	yes
Grinding	yes	no	no	no
Shear Cutting	yes	yes	yes	yes

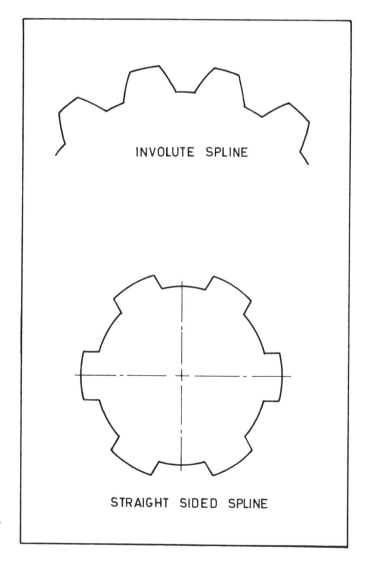

Figure 6-2. Tooth forms of involute splines and straight sided splines.

Tooth spacing tolerances and runout are less critical than with gears. As a result a wider variety of processes is available for spline production and more than one method can be used to produce a specific spline. The fact that a particular machine is available on the shop floor is surely a very important argument in selecting the method. There are however a number of preferred methods which are universally used in the industry.

In Table 6-2, a preferred method is indicated, together with alternate methods which can be employed depending on availability of equipment and production quantities.

Production Volume

The output of an operation must meet the production schedules either by the inherent speed of the machine or by multiple machine installations performing identical operations. Equipment for production of splines is offered in many versions which enable low-volume and high-volume manufacturers to satisfy their specific needs in the most economical way.

Production machines for splines are built according to one of the following principles.

1. Generating machines with basic rack or involute profile tools. Machines: hobbers, shapers, spline rollers.
2. Cutting one tooth at a time with full form tools. Machines: spline millers, grinders.
3. Simultaneous cutting of all teeth with full form tools. Machines: internal broaches, potbroaches, shear speeders, progressive broaches.

The first category of machines has the flexibility to handle both low and high-volume production. Relative to speed, the rolling process is capable of meeting the outputs required for high-volume production with a single machine. Hobbing and shaping are slower and require multiple machine installations in high-volume lines.

Spline millers and spline grinders which machine one tooth at the time are obviously low output machines which find their application mainly in batch production. The output of spline millers can, however, be increased on multispindle machines which mill two or three parts in the same cycle.

The third category of machines find their application in high-volume lines with the exception of internal spline broaching which is used also for batch production. External broaches or potbroaches, shear speeders, and in particular, progressive broaches are almost exclusively used for high-volume production.

SPLINE MANUFACTURING PROCESSES

Spline Rolling

Spline rolling is accomplished by rolling a solid blank between two basic racks profiles which form the tooth by plastic deformation of metal. The diameter of the shaft before rolling is chosen in such a way that the amount of material displaced below the diameter equals the amount of material above.

Table 6-2
Spline Manufacturing Processes

Spline	Tooth geometry	Preferred Method	Alternate Method
Type A	Involute P.A. = 45 degr.	Rolling	Hobbing Milling Shaping Potbroaching Shear Cutting
	Involute P.A. < 45 degr.	Hobbing	Milling Shaping Potbroaching Shear Cutting
	Straight sided	Milling	Shaping Potbroaching Hobbing Shear Cutting
Type B	Involute P.A. = 45 degr.	Rolling	Shaping Shear cutting Progressive Blind Broaching
	Involute P.A. < 45 degr.	Shaping	Shear cutting Progressive Blind Broaching
	Straight sided	Shaping	Shear cutting Progressive Blind Broaching
Type C	Involute P.A. = 45 degr.	Broaching	Shaping Shear Cutting
	Involute P.A. < 45 degr.	Broaching	Shaping Shear Cutting
	Straight sided	Broaching	Shaping Shear Cutting
Type D	Involute P.A. = 45 degr.	Shaping	Progressive Blind Broaching Shear Cutting
	Involute P.A. < 45 degr.	Shaping	Progressive Blind Broaching Shear Cutting
	Involute P.A. < 45 degr.	Shaping	Progressive Blind Broaching Shear Cutting

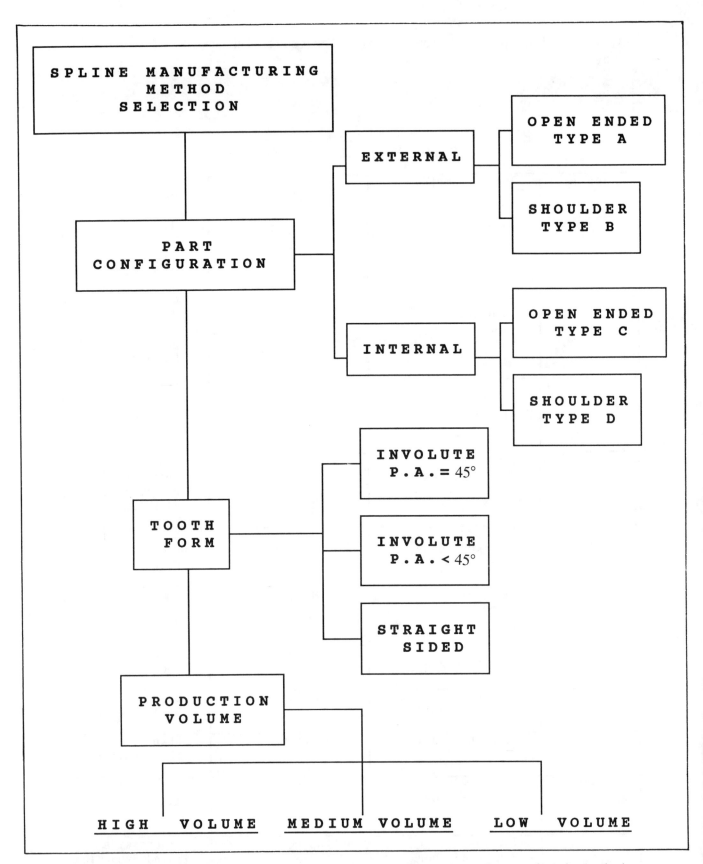

Figure 6-3. Decision tree diagram for the selection of the most appropriate spline manufacturing method.

For the best results, the blank diameter tolerance must be held within 0.002 in. (0.05 mm) or less, and must have minimum ovality or taper. The runout of the blank diameter, relative to the centers, must also not vary more than 0.002 in. (0.05 mm). Because of the accuracy required for successful rolling, the diameter is often ground to size prior to rolling.

The Roto-Flo rolling process developed by Michigan Tool Company is a fast and economical method of producing involute tooth forms. Many types of splines, threads, grooves, or serrated forms, can be rolled on the machine in cycle times ranging from 10 to 14 seconds. Rolling rates on machines equipped with automatic part loaders may range from 240 to 350 pieces per hour at 100% efficiency.

The forming tools, or racks, are designed in such a way that the tooth height increases toward the end of the rack. To ensure close tolerance control and smooth finish, a number of teeth at the finish end are of the same height to allow several revolutions of the work while at full depth contact with the tool.

Rolling is a typical high-speed, high-volume, production method for the automotive industry. It produces splines with smooth surface finish, and high strength due to the cold working of the material and superior grain structure. The hardness and ductility of the material must be controlled within close limits for consistent quality results.

Splines with pressure angles between 30 and 45 degrees can be successfully rolled. However the process is most economical with pressure angles of 45 degrees. The two main factors that affect tool life are spline pressure angle and material hardness. Tool life may be as high as 150,000 pieces between regrinds for 45 degree pressure angle splines, and as low as 50,000 pieces for 30 degree.

Because of tool cost, the Roto-Flo rolling process is used only for involute splines with 45 degree pressure angle while other methods are used for smaller pressure angle splines.

Spline Milling

This process uses a form milling cutter to cut one tooth space at a time. An indexing head rotates the work to the next tooth space after each cut, until all teeth are cut.

Spline milling is a universal process capable of producing involute and straight sided spline, and other tooth forms as well. To increase productivity, special spline milling machines are built with two or three spindles capable of producing two or three parts per cycle. Spline milling is often used to produce straight sided splines on shafts.

Internal Spline Broaching

This method of broaching is the most efficient method of cutting open-ended internal splines in parts such as gears, hubs, couplers, connectors, etc. Involute splines, straight sided splines, serrations as well as any other type of tooth form can be cut by broaching.

In internal broaching the tool is pulled through the part and the spline is cut to full depth in a single pass.

Concentricity broaches are a special version of spline broaches with a finishing segment consisting of alternate round shells and full form involute teeth. The round finishing shells cut the minor diameter of the spline in the same cycle as the spline teeth. The result is a smoothly finished minor diameter that is held within close diameter tolerance and almost perfect concentricity (within 0.0002 in. or 0.005 mm) with the pitch diameter of the spline.

Processing parts with an internal spline is simplified if the minor diameter is precisely concentric with the pitch diameter of the involute teeth. This permits the use of solid round arbors in subsequent operations with assurance that concentricity between spline and external features is maintained.

External Spline Broaching

External spline broaching, also called potbroaching, is used to cut involute and special tooth forms or grooves on the outside diameter of a part.

The part is either pulled or pushed through the stationary broaching tool, and the spline is cut in a single pass. Tools are dedicated to one part, making the operation only feasible for high-volume applications. Potbroaching machines can be either horizontal or vertical. The process is only applicable to open-ended external tooth forms.

Progressive Blind Broaching

Progressive broaching, or transfer broaching, is a method developed to cut internal or external blind splines. A progressive broach is a transfer type machine in which the part is moved automatically through a number of broaching stations. Each station is equipped with a broaching tool which removes a specific amount of material in a single stroke. The size of the broaching tool increases from station to station until the spline tooth is finished to final size in the last station. Progressive broaching is a high-speed spline cutting method that can only be economically applied for high-volume work.

Shear Cutting

Shear cutting, or shear speed, is a process that uses a reciprocating cutting stroke similar to shaping to cut all teeth simultaneously, but without the rotating motion of the part which is typical for shaping.

The cutter head assembly remains stationary during operation and the work holding fixture reciprocates the part vertically through the cutter blades. Shear speed blades, precision ground to form the spaces between the teeth of the spline, are advanced radially with each stroke until full depth is reached.

The advantage of shear speed is that it can cut any internal or external spur tooth form in much shorter time than conventional shaping. The presence of an adjacent shoulder is no problem for shear speeding.

Accuracy in index and spacing is, however, determined by the manufacturing quality of the cutter head, spacing errors are duplicated in the work pieces. This is why shear cutting of gears is only rarely done.

The process is not flexible. Cutter heads are designed to make one particular spline. Parts with critical tolerances represent a real challenge because of the inherent spacing errors in the tooling.

Spline Hobbing

Spline hobbing is an operation which is routinely applied in manufacturing of drivelines for automotive, truck, agricultural, and construction equipment. Because of the stub tooth design, the cutting depth for splines is less than for gears and a single hobbing pass is, in general, sufficient to cut the spline to finish size.

Spline hobbing is performed at reduced feedrates compared to gear hobbing, to limit the size of the feed marks and to produce an acceptable surface finish. Mating splines which slide axially over each other in assembly, as used in sliding couplers, may require a shaving operation after hobbing to provide a smooth surface which facilitates the sliding action.

Hobbing is the preferred method for open-ended involute splines with 30 degrees pressure angle. Forty-five degrees pressure angle splines are produced more economically with the rolling method. Straight sided splines can also be hobbed with a cycloid tooth form hob.

Spline Shaping

Spline shaping is also a very common operation in driveline manufacturing, typically applied for external and internal involute splines with an adjacent shoulder, and with pressure angles of 30 degrees. Compared to hobbing, it has the advantage of producing a very smooth surface finish because the tool marks run parallel with the axis of the part. Involute splines are shaper cut in the continuous generating method, straight sided splines are cut on shapers with automatic indexing mechanisms.

Spline Shaving

This is a process used only in special cases to improve surface finish of hobbed splines when the splined members must slide in axial direction, or to provide a crowned or tapered tooth form.

Spline Grinding

Splines are ground when high importance is attached to accuracy and durability in power transmission. Splines are ground one toothspace at the time with a full form grinding wheel (Figures 6-4 and 6-5). Open-ended

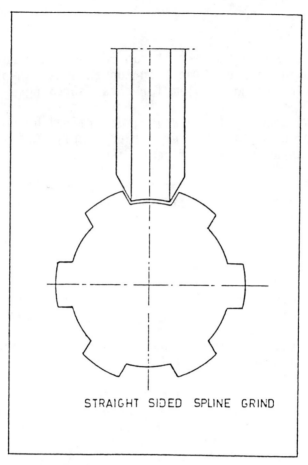

Figure 6-4. Grinding of straight sided splines, one tooth space at a time, with a form grinding wheel.

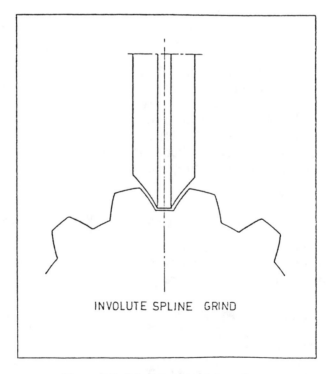

Figure 6-5. Grinding of involute splines, one tooth space at the time, with a form grinding wheel.

splines can be ground in a cycle that grinds on the forward pass with a spark out on the return. The workpiece is then indexed and the sequence is repeated for the required number of splines. Infeed is applied after each completed revolution of work.

Accuracy of spacing is controlled by the use of precision index plates. Automatic wheel dressing devices form the grinding wheel in the proper shape for involute and straight sided splines and serrations. Involute splines can also be ground with the continuous generating method developed by Reishauer.

7
Special Gear and Spline Processes

In this chapter, three specific operations are discussed. The three are often encountered in gear process plans, but are not directly related to manufacturing processes for gear and spline teeth.

GEAR CHAMFERING AND DEBURRING

Chamfering and deburring of gear teeth has a significant impact on the sound level performance of gears. Gear noise is, in many instances, caused by tooth profile imperfections which impede smooth engagement of mating gear profiles. The cause of these profile imperfections can be traced back to two possible sources.

* Burrs and sharp edges formed in hobbing and shaping.
* Nicks and other damages resulting from careless handling of gears after the teeth have been cut.

Heat treatment, the converging point for all gears produced in a manufacturing plant, is a likely source of damage because of the large quantities of parts handled in bulk through heat treatment furnaces.

In hobbing and shaping, burrs are formed on the end faces of the gear teeth. The size and severeness of burrs increases with progressing wear of the cutters employed.

Besides functional problems in assembly, burrs can also cause injuries in manual handling of parts, and quality problems in subsequent gear finishing operations, like shaving.

Special gear deburring machines are available which eliminate the burrs by cutting a chamfer along the edge of the gear tooth. (Figure 7-1) The chamfer can be cut along the complete tooth profile, or partially along the working depth of the involute. The machines can handle both spur and helical gears, and chamfers can be generated on both flanks of the same tooth, or on one flank only as is sometimes specified for the acute angle of helical gears.

The cutting tools used on chamfering machines are either high-speed steel cutters or grinding disks. High-speed steel cutters can only be used before heat treatment, and the chamfering operation is performed between hobbing and shaving or between shaping and shaving. Machines equipped with grinding wheels can be employed before or after heat treatment. A chamfering operation after heat treatment has the advantage of eliminating handling

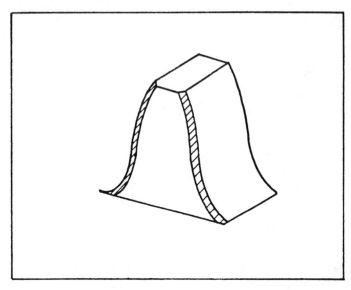

Figure 7-1. Tooth chamfering as applied for gears and splines.

damages in addition to effective elimination of burrs resulting from gear cutting processes.

TOOTH ROUNDING AND TOOTH POINTING

Tooth rounding and tooth pointing are operations performed on spline teeth of transmission gears to facilitate the engagement of a sliding coupler. In manually shifted constant mesh transmissions, gears are engaged and disengaged by moving a sliding coupler with an internal spline over the external spline of a gear. Once engaged, the sliding coupler functions as a solid connector between gear and shaft. Because of the small amount of free play between the mating splines of gear and coupler, and because of the relative motion between rotating parts in a transmission, engagement would not be possible without a lead-in feature at the end of the spline. For this purpose, special tooth chamfering machines have been developed to generate lead-in chamfers or other special shapes at the end of spline teeth.

Tooth rounding and pointing machines are equipped with a special form cam which controls the movement of the cutter in relationship with the rotation of the workpiece. The operation is completed in one full rotation of the workpiece in a short cycle time. Internal and external splines can be chamfered on the same machine.

The two most common operations performed on gears with dogtooth splines are tooth rounding and tooth pointing. (Figure 7-2) Tooth rounding is applied for nonsynchronized gears used in tractor transmissions and consists of cutting a radius at the ends of the spline teeth. Tooth pointing creates a pointed tooth form with straight sides and is universally used for synchromesh gears in tractor, truck, and automotive transmissions. Single flank chamfering provides smooth engagement of pinion and starter ring gears.

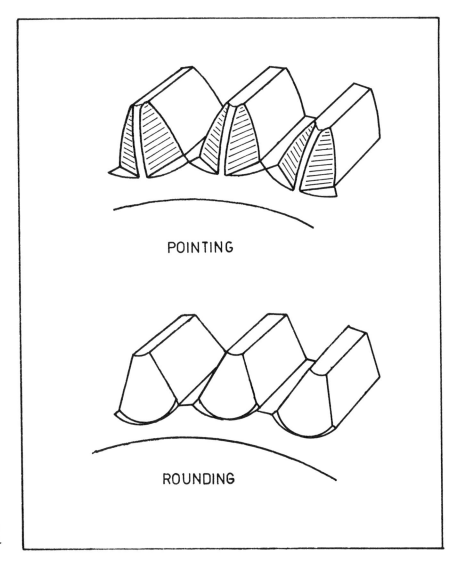

Figure 7-2. Pointing and rounding of spline teeth as applied for coupler splines.

GEAR SOUNDING

Gear sounding is a testing method used to detect noisy gears prior to assembly. By identifying defective gears before they are assembled, corrective actions can be taken without incurring expensive and time consuming teardowns of transmissions which fail to meet sound performance criteria.

In a gear sound tester, the work gear is run together with a mating gear in a specially designed acoustical sound box that amplifies the sound through a horn. Abnormal gear noise is easily detected by the operator who can immediately proceed with corrective action. A special sound proof room is not required.

Low-speed gears do not require matching and can be run with a master gear through the sound tester. High-speed gears are normally matched in sets after the sound test and are transported as sets to the assembly line.

Sounding is very effective in detecting knocks produced by irregularities in the tooth profile. Knocking caused by nicks and burrs can be corrected immediately by smoothing the defective areas with an abrasive stone. Other types of gear noise which are the result of variations in profile, eccentricity, index, and other dimensional errors can be corrected by gear honing.

To eliminate the human factor in evaluating noise , gear sound testers have been developed equipped with electronic noise level indicator which allow testing gears against a preset noise level. It is also possible to equip the machine with an automatic electro deburring device which will remove small nicks and burrs during the test cycle until the sound specification is met.

8
Blanking

SCOPE

The term "blanking" refers to the initial metalcutting operations in the process planning sequence which produce the contour of a part starting from rough material.

The scope of blanking is:

* To remove the excess material.
* To machine the part to print specifications, except for those surfaces with subsequent finishing operations.
* To leave adequate machining stock for finishing operations
* To prepare good quality surfaces for location and clamping of the part throughout the process.

Process Selection

Processes and machines for blanking of driveline parts are selected based on:

* part configuration,
* production volume.

Part Configuration

Gears, countershafts, and clustergears can be described as being rotational parts. As such, the contour of the part can be generated by a single point tool travelling longitudinally along the axis of the part, while the part rotates around its axis.

Depending on the diameter-to-length ratio, rotational parts can be further classified in disk type and shaft type parts (Figure 8-1).

The sequence of operations and machinery required to produce blanks is characteristic for each basic configuration.

Disk type parts, such as gears, are turned on open-ended lathes or chuckers which locate and clamp the part on internal or external diameters in concentric chucks.

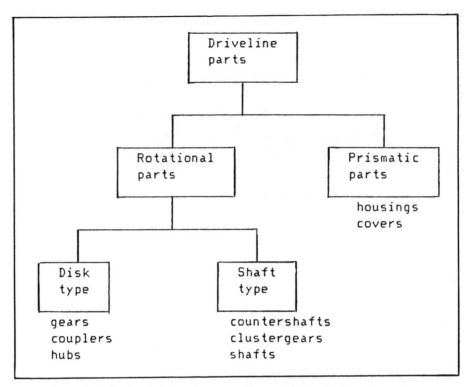

Figure 8-1. Basic classification of parts according to their configuration.

Shafts, on the other hand, are located between centers. The first operation consists in machining the end faces of the shaft to a specific length, and drilling opposite and in-line center holes.

Production Volume

The type of machine used for a specific blanking operation and the degree of automation is determined by the required output in pieces per hour of the operation.

Low-volume production is processed on manually operated, universal machinery, with standard low-cost tooling. This type of job shop equipment typically handles a wide variety of parts in lot sizes of one to several hundred.

A medium-volume production line is set up to handle a family of 10 to 20 parts in quantities of several thousand per year. Monthly requirements are produced in batches of 500 to a few thousand, with line change-over between production runs.

This is the sector of the manufacturing industry where CNC machines are employed to their fullest potential. The versatility and change-over flexibility of CNC equipment make it ideally suited for short intermittent production runs. The universal tooling packages used on CNC lathes and chuckers, combined with short change-over time, result in efficient batch production at low inventory level.

The inherent accuracy of CNC machines offers the added advantage of roughing and finishing in the same setup, reducing the number of machining operations in the process plan.

Characteristically for a high-volume production line, machining operations must be completed in very short cycle times. For instance, to make one million pieces per year in a three-shift operating pattern at 80% efficiency, machine cycle time plus part handling must be less than 15 seconds per part.

To meet line speed demands, roughing and finishing operations are processed on separate machines. Multispindle machines with special dedicated tooling are used to cut more than one part per cycle. Automatic loading and unloading of parts is used extensively to reduce idle time to the minimum.

Rough Material

Barstock, forgings, extrusions, or precision forgings are used to make driveline gears and shafts. The most suitable raw material form is selected based on a financial evaluation which includes all manufacturing costs, raw material, vendor tooling, in-house machining cost, and investment in facilities and tooling.

Barstock is the simplest and cheapest form of raw material frequently used for small-lot sizes. More complex fabricated forms like forgings and extrusions become economically attractive with increasing production volumes.

BARSTOCK

Hot rolled barstock is extensively used in job shop work. The advantages of barstock are:

* Low purchase cost. Rolled barstock from the mill is the most inexpensive material available.
* Short lead time. Barstock is offered in standard sizes which are available from stock.
* No vendor tooling cost. Barstock is selected from existing standard sizes and requires no special operations by the vendor.
* Many sources of purchase. Changing from one supplier to another does not create problems. Purchasing is not locked in with only one supplier.

Disadvantages of Barstock

The disadvantages are:

* High material waste. Parts with large variation in diameter have a high percentage of wasted material.
* Long machining time. To remove excess material many turning passes are required on lathes.
* More machining operations. Barstock needs to be cut to individual lengths before it can be processed.
* Special material handling for transport and storage.
* Material strength may be affected due to the interrupted fiber flow in the part.

Table 8-1
Process, Machines, and Equipment,
Characteristics in Low, Medium, and High-Volume Manufacturing Operations.

Process Characteristics	Low Volume	Medium Volume	High Volume
Machines and Equipment			
Universal Machines	yes	yes	no
CNC Machines	yes	yes	no
Multispindle Machines	no	no	yes
Special Machines	no	no	yes
Manual Operated	yes	no	no
Automatic Cycle	no	yes	yes
Material Handling			
Manual Part Handling	yes	yes	no
Automated Part Handling	no	no	yes
Flexible Automation - Robots	no	yes	yes
Tooling			
Universal Tooling	yes	yes	no
Special Tooling	no	no	yes
Operating Conditions			
Change Over	yes	yes	no
Number of Different Parts	>100	10-20	1
Labor Content	High	Low	Low
Floor Layout	Stand alone	Line	Line
Retooling Cost for New Parts	Low	Low	High
Production Lot Sizes	1-500	500-2000	Continuous
Annual Production Volumes	1-500	500-20,000	>100,000

In general, there is no cost advantage in using barstock instead of fabricated raw material. It is the appropriate solution for low-volume production because it avoids investment in vendor tooling, and is readily available. The low material cost is, however, more than offset by the cost of additional machining, and the high proportion of wasted material.

When the variation in section is small, barstock can be used economically for high-volume production. Examples are screw machine parts, shifter rails, spool valves, pins, etc.

FORGINGS

Forgings are used for medium and high-volume production. The fabricated form is made as close as possible to the finish contour of the part to reduce machining time, and to keep the amount of wasted material low. Forging tolerances are held within ± 0.030 in. (0.8 mm) and machining stock allowance is in the range of 0.080 to 0.120 per side. The surface finish of forged surfaces is around 500 microinches.

Advantages of forgings

Advantages are:

* Short machining times. The raw material is shaped close to the final contour of the part and excess material can usually be removed in a single turning pass.
* Less machining. Nonfunctional surfaces do not require machining. Forging tolerances and surface conditions permit usage of parts with surfaces in forged condition.
* Higher material strength due to the oriented fibers and the compact material structure.
* Easy to handle.
* Easy to transport.

Disadvantages

Disadvantages include:

* Forgings have parting lines, draft angles of 3 to 6 degrees, and flash trim marks which affect tooling performance.
* Higher material cost in dollars per unit of weight.
* Forgings require vendor tooling and lead time.

Forgings are economically feasible even at relatively low volumes of 500 to 1,000 pieces per year. Vendor tooling cost for drop forgings and upset forgings commonly used for gears and shafts can be written off over a few thousand pieces.

Precision hot forgings, or flashless forgings for automotive parts are produced on transfer forging presses. Precision forgings have minimum draft angle of 1/2 to 1 degree and no parting lines. Machining stock allowance is only 0.040 in. (1.0 mm) and length and diameter dimensions are held within ± 0.012 in. (0.3 mm) tolerance.

EXTRUSIONS

Cold extrusion is the most sophisticated form of rough material which finds economic application in high-volume production of automotive parts.

Advantages of Extruded Parts

The advantages of extruded parts are:

* Close tolerance control. Extrusion tolerances can often be held within 0.005 in. (0.13 mm) on the diameter and 0.060 (1.5 mm) overall length.
* Excellent surface finish. Cold extrusion produces surfaces with finishes ranging between 32 and 125 microinches (0.8 and 3 mm).
* Improved mechanical properties. The work hardening effect of cold forming increases the tensile strength and yield strength.
* Minimum wasted material because of the close tolerances which can be held with extrusion techniques.
* Machining savings. Extruded surfaces may be used in the final product because of the extremely smooth finish and close tolerance control.

Disadvantages

The disadvantages are:

* High piece cost.
* Very high vendor tooling cost. Investment in equipment and tooling is virtually never justified by low and medium volume production requirements.

EXAMPLES OF BLANKING OPERATIONS

The following three examples of blanking operations illustrate the principles explained in this section.

Example 1

A small planet gear which, because of the small outside diameter and the uniformity in section, is made most economically from barstock (Figure 8-2).

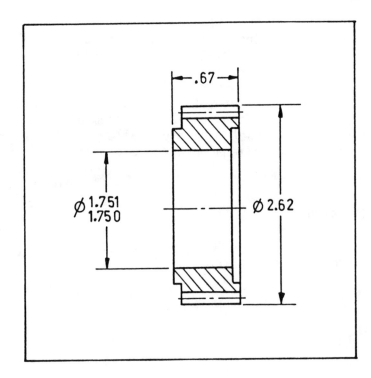

Figure 8-2. Planet gear made from barstock.

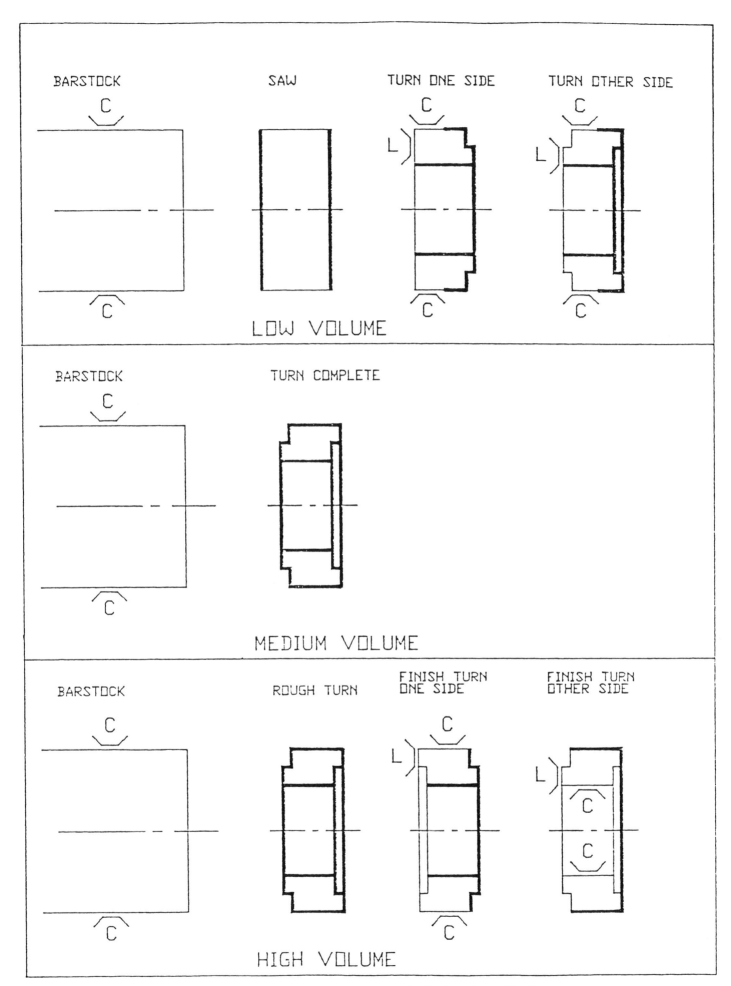

Figure 8-3. Process planning sequence for turning of gear blanks.

When a small quantity of parts is required, as for prototype work, the barstock is sawed into individual pieces, and the part is turned completely in two setups on a lathe (Figure 8-3).

Short intermittent runs of 500 to 2,000 pieces are produced most effectively on a single spindle CNC barchucker that performs the complete rough and finish turning operation in one setup.

A mass production manufacturer will utilize a multispindle barchucker to perform the rough turning operation, followed by two separate setups to finish turn both sides of the gear.

Example 2

Transmission gear blank made from barstock for low-volume, and from forging for medium and high volume (Figure 8-4).

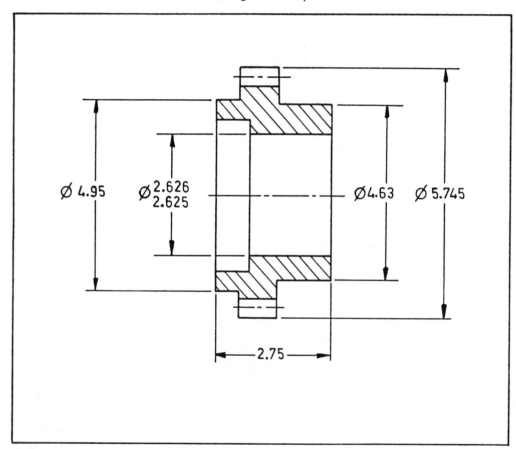

Figure 8-4. Transmission gear blank made from barstock for low-volume, and from forging for medium and high-volume.

The low-volume process consists of cutting the barstock to length and roughing and finishing in two setups on a lathe (Figure 8-5).

Medium-volume processing starts with a forging which is rough and finish turned in two setups on a single spindle CNC chucker.

To meet the required output per machine, the high-volume process splits the turning operations in separate roughing and finishing setups on multispindle machines.

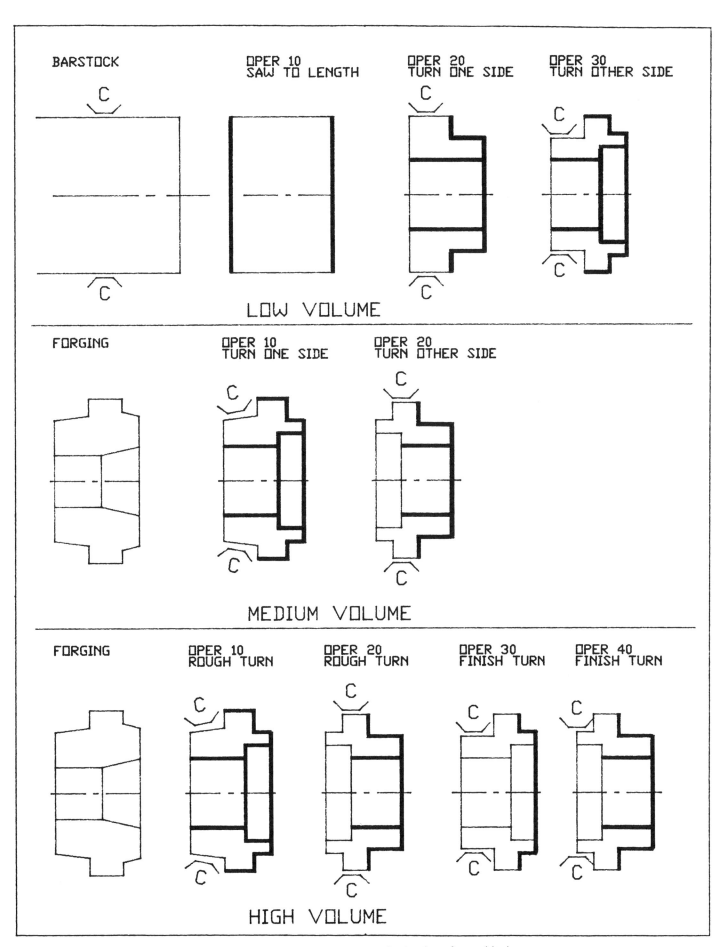

Figure 8-5. Process planning sequence for turning of gear blanks.

Example 3

Figure 8-6 illustrates a transmission counter shaft with a gear on one end of the shaft.

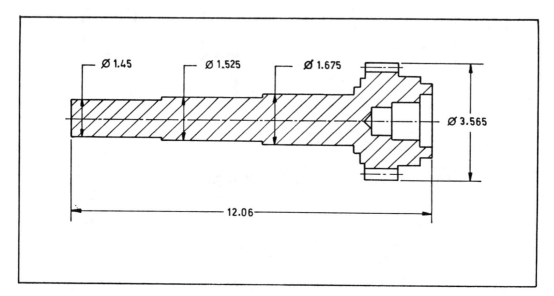

Figure 8-6. Transmission countershaft with a gear on one end of the shaft.

The low-volume process starts with barstock which is cut to length and prepared for turning by facing both ends and drilling center holes (Figure 8-7). Two turning operations are planned to finish turn the external contour of the part. This is followed by an internal turning operation on a chucker where the part is clamped and located on the outside diameter near the gear. This example illustrates the inefficient material utilization of barstock as almost half the material is wasted in chip removal.

The high-volume process starts with a forging which is faced and centered on both sides while the internal diameters are roughed in the same setup. Two copy turning operations turn the external contour. The outside diameter near the gear is ground in operation 40 to control the runout of the diameter used as locator in operation 50 (Figure 8-7), and in the gear cutting operation which is not shown here.

A medium-volume process with fewer operations is possible by combining operation 30 and 50 in one setup. The external and internal diameters are rough and finish turned on a CNC chucker in operation 30. Grinding of the location diameter is not required provided good runout tolerance can be maintained in operation 20. If operation 20 also is performed on a CNC lathe the location diameter can be machined to close tolerances in one additional finishing pass (Figure 8-8).

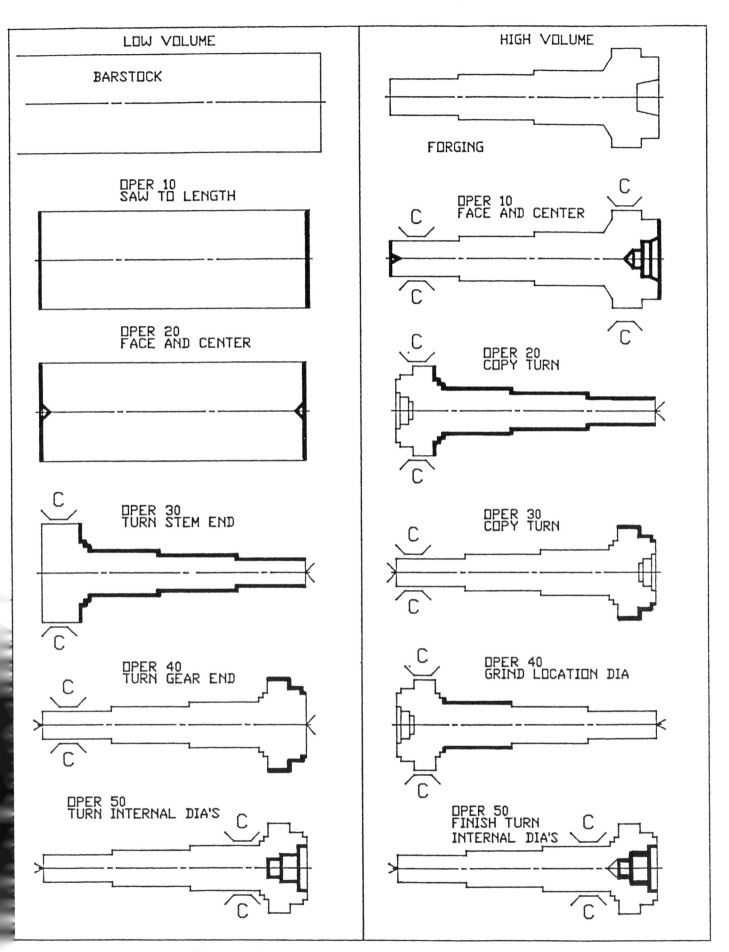

Figure 8-7. Process planning sequence for blanking of shafts.

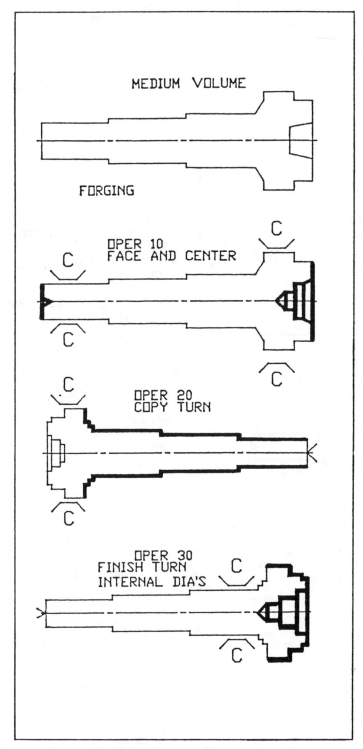

Figure 8-8. Process for medium volume.

9
Heat Treatment

High loaded gears are heat-treated to create a material structure and mechanical properties that increase wear resistance, fatigue life, and impact resistance. These properties greatly improve the performance of gears and are sought after by design engineers. Process planners must, however, contend with the negative side effects of heat treatment. This negative side includes a drastic reduction in machinability, and a quality deterioration which is difficult to control.

An in-depth discussion of heat treatment methods is beyond the scope of this book. It will, however, cover briefly the various heat treatment processes currently applied for driveline parts.

HEAT TREATMENT PROCESSES

There are basically two categories of heat treatment processes.

* Processes that produce material properties favorable for machining.
* Processes that produce high strength and surface hardness after the parts have been machined.

The first category of heat treatments is executed after fabrication of the rough material at the forging shop, and before the material is received in the manufacturing plant.

The second category is usually performed in the manufacturing plant after full or partial machining of the parts. In plant heat treatment is an integral part of the process planning sequence of driveline components.

Heat treatment of machined parts can be further divided into case hardening methods and through hardening methods.

Case hardening creates a material structure which is hard and wear resistant at the surface and tough, soft, and impact resistant at the core.

Through hardening of medium carbon alloy steel creates a uniform structure throughout the material with high-strength and impact resistance.

Two methods of case hardening are currently used:

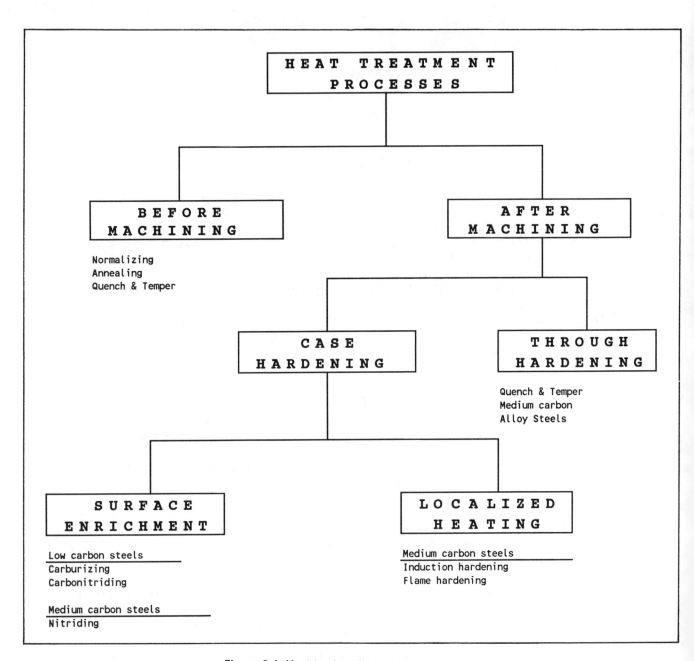

Figure 9-1. Heat treatment processes.

* By enrichment of the surface of low carbon steels with a hardness causing element as in carburizing, or carbonitriding. Or by nitriding of medium carbon steels.
* By localized heating and quenching of medium carbon steels as in induction hardening and flame hardening.

Normalizing

This heat treatment process is performed after hot working operations like hammering, rolling, forging, or extrusion that shape the metal into the desired form. Normalizing consists in reheating the material to a temperature of 1500 to 1600 degrees F (815 - 871.1 degrees C) and slow cooling in air. The purpose of normalizing is:

* To refine a coarse grain structure resulting from hot working operations at high temperature.
* To create an even and fine distribution of carbides in steel which results in a slight increase in hardness, a more brittle chip, and better machinability.
* In alloy steel, the finer carbide particle size obtained by normalizing facilitates subsequent heat treatment resulting in a more uniform microstructure.

Annealing

Annealing is a controlled heating and cooling process executed with the purpose of:

* Softening of the material
* Relieve cooling stresses resulting from hot working.
* Improve machinability.
* Improve the material's ability to be subsequently formed.

Full annealing consists of heating steel to 1500 - 1600 degrees F (815 - 871.1 degrees C) followed by slow cooling in the furnace. In low and medium carbon steel, this simple procedure creates a coarse pearlite structure. This structure is good for machining.

In isothermal annealing, the cooling rate in the furnace is controlled. The steel is held at an elevated temperature until the pearlite transformation is complete. The steel is then rapidly cooled to room temperature. Isothermal annealing is faster than full annealing.

Spheroidize annealing is a heat treatment performed to improve the machinability of high-carbon steels. The steel is held at high temperature in the furnace until the carbides are completely transformed in spheroidal particles. The coarse pearlite structure is too hard in high-carbon steels, and machinability is improved with a globular carbide structure.

Quench and Temper

Quench and temper of medium carbon steels, when executed prior to machining, can entirely eliminate the need for subsequent heat treatment resulting in lower manufacturing cost.

The cycle consists of heating steel to hardening temperature and quenching in oil, followed by a reheat to a temperature high enough to reduce the material hardness to 30 - 35 Rc. A tough, impact resistant structure is obtained which still maintains good machinability. The decrease in tool life experienced with this kind of material is more than offset by the elimination of heat treatment and subsequent grinding operations.

Quench and temper after machining consists of heating medium carbon alloy steel to hardening temperature and quenching in oil, followed by reheating to a temperature of 300 to 400 degrees F (148.9 - 204.4 degrees C) to soften and toughen the steel. The low temperature reheat causes a slight reduction in hardness (56 - 58 Rc), but increases impact resistance and relieves quenching strains.

Carburizing

Heating a low-carbon steel to a temperature of 1600 - 1650 degrees F (871.1 - 898.8 degrees C) in presence of a gaseous or liquid carbonizing atmosphere is known as carburizing. Under prolonged exposure, carbon penetrates and diffuses into the material creating a thin layer of high-carbon steel. The thickness of the carbon layer is directly related to the exposure time, while the penetration rate can be increased by raising the temperature.

Quenching in hot oil causes the carburized case to become very hard, while the low carbon core remains soft and tough. Surface hardness of carburized parts ranges from 58 to 62 Rc.

Liquid carburizing consists in immersing the parts in a molten salt bath which contains the carburizing agent. Gas carburizing is mostly used in medium and high-volume production and is a process that lends itself well to continuous heat treatment.

The deep case carburizing method is used for parts with case depth varying from 0.020 to 0.250 in. (0.5 - 6.3 mm), while light case carburizing is used for parts with 0.035 maximum case depth.

Carbonitriding

This process differs only from carburizing in that nitrogen is also present in the atmosphere besides carbon. The nitrogen penetrates and diffuses into the material and combines with the alloy elements to form very hard nitrides resulting in increased surface hardness. The surface hardness range obtained with carbonitriding is 60 - 65 Rc.

Nitriding

Nitriding is a nitrogen adding case hardening process for finish machined parts with very low distortion effect, eliminating the need for subsequent machining. Nitrided parts have a very wear resistant surface and maintain their hardness at elevated temperature.

In nitriding, parts are heated to a temperature of 950 to 1000 degrees F (510 - 538 degrees C) in the presence of ammonia gas, the nitriding agent. Because nitriding temperature is low, no structure transformations occur and distortions are minimum. The nitriding case is very thin, 0.004 to 0.015 in. (0.1 - 0.4 mm), and requires much longer furnace time than carburizing. The case is much harder than with carburizing and ranges from 64 to 78 Rc.

Aluminum and chromium are alloy elements which facilitate the nitriding process. Core strength is not affected by the process. To provide sufficient core strength, parts can be quenched and tempered prior to nitriding.

Nitriding is an expensive process with the disadvantage of making the case too thin to withstand high contact pressure.

Induction Hardening

Localized heating and quenching by means of high-frequency currents induced in the surface of the parts is called induction hardening. Medium carbon steels with carbon content between 0.4 - 0.6% are used for this process.

The heating effect is confined to a limited area, controlled by the shape of the induction coil. The power input per unit area is so high the temperature of the workpiece is raised rapidly to hardening temperature. Immediate quench prevents heat loss to the underlying material.

To harden a large area, it is necessary to traverse the coil over the face of the metal to be hardened. The case depth is controlled by the power input and the traversing speed.

Quenching can be by water, oil, or other quenching medium and is applied by complete submersion or spray quenching.

Because of the short cycle times induction hardening is suitable for high-volume production. The machines are environment friendly and can be installed in the normal machining line. Induction hardening is normally used for case depths ranging from 0.020 to 0.140 in. (0.5 - 3.5 mm).

Flame Hardening

Similar to induction hardening, flame hardening consists of localized heating and quenching of parts by means of a high temperature flame. The process is also used with medium carbon steels. It is a slower and less accurate method than induction hardening which is not used in high-volume production.

HEAT TREATMENT DISTORTIONS

There are two main reasons for heat treatment distortions.

* Residual internal strains resulting from the fabrication method used to make the rough part. Hot and cold forming processes like forging, and extrusion displace metal under very high pressure and create a

specific pattern of fibers in the internal structure. Plastic deformation leaves high internal stresses in the material which are released in heat treatment and result in warp and distortion.

* Heat treatment in itself is a major cause of unbalanced strains. Parts with uneven mass distribution are subject to different cooling rates during quench resulting in tension and compression strains which distort the part.

Since distortion is an inherent and inevitable result of heat treatment, the process plan must include operations and procedures such as those outlined in the following discussion to correct or minimize the amount of distortion.

Restore Geometry by Plastic Deformation

A distorted part can be brought back to its original form by exerting mechanical pressure on the part in the opposite direction of the distortion.

Flex-bending, or overbending straightens shafts on hydraulic presses. The high point of the bow is found with a dial indicator and a force is applied to push the bowed area back beyond the elasticity point so it returns to a straight position when the force is removed.

Peen straightening is another method of straightening shafts which consists in peening, or hammering of the low side of the bow opposite the high point. The hammering action displaces material on the surface and creates compressive stresses which compensate for the bending stresses.

Out-of-round conditions of thin walled ring gears can be corrected on a roll rounding machine. The internal ring gear is meshed with a pair of external gears, one of which rotates the ring gear while the other moves outward and exerts pressure. Up to 75% of the unroundness can be eliminated with this method.

Eliminate Distortion by Machining After Heat Treatment

This is, of course, the most effective way of dealing with heat treatment distortions, but it is expensive and the objective of efficient process planning is to keep the amount of machining after hardening to a minimum. Internal and external grinding, surface grinding, hard finishing of gears, lapping and honing are the most common manufacturing processes.

Nonabrasive machining methods require adequate machinability which is no longer present in hardened gears. Case carburized parts can be sprayed with a carburization inhibitor to prevent hardening of selected surfaces. These surfaces maintain sufficient machinability to be machined with a conventional cutting process. This method is applied to rebroach internal splines, and to drill and ream holes after heat treatment.

Compensate for Heat Treatment Distortion

This is a procedure often applied for precision parts and referred to as heat treatment development. The technique consists of measuring the amount and direction of distortion of a particular feature on a sufficiently large number of parts so that a statistical evaluation can be made of the average dimensional change. The machining tolerances before heat treatment are then modified with the same amount but in opposite direction.

This method compensates for lead unwind in helical gears, and for involute profile changes. The change in helix angle can easily be compensated by cutting the gear according to a corrected helix angle. Changes in involute profile are compensated by grinding modifications in the profile of the generating tools. Properly executed, heat treatment development can avoid expensive hard finishing of gears.

Changes in linear and diametral dimensions can be compensated in the same way. A good example is changing the dimension over pins of involute splines to compensate for heat treatment variation.

Low Distortion Methods

Some heat treatments offer the advantage of low distortion and can be used for intricate parts which cannot be handled by other processes.

Nitriding is a low-temperature heat treatment which increases surface hardness with almost no distortion. It is, however, a time consuming and expensive method.

Press quenching is commonly used in driveline manufacturing for rear axle bevel gears and other parts which are susceptible to warp.

Induction hardening is also frequently used for localized hardening leaving the rest of the part soft and not affected by distortions.

Quench and temper before machining, if permitted by the part function, can be used to make good quality parts without subsequent heat treatment. For adequate machinability the maximum hardness should not be higher than 38 - 40 Rockwell C.

Design Practice

Good design practice can contribute a lot toward heat treatment feasibility. Distortion and warp can be minimized with part configurations which have the following features:

    * no large variation in section,
    * are not asymetric,
    * equally distributed mass,
    * no extreme length to diameter ratio,
    * no abnormally thin sections.

Obtaining a uniform case depth in carburized parts is heavily influenced by design. The penetration depth of carbon will vary in parts with unequal mass. Thin sections will reach carburization temperature faster than thick sections and will be exposed longer to carbon penetration and diffusion resulting in deeper case.

Case depth is also less in internal surfaces which are not freely exposed to the carbon atmosphere. Bores in gears have typically less case depth than external diameters. The problem is exacerbated by grinding operations which further reduce the case depth (Figure 9-2).

Figure 9-2 shows a part design identifying several case depth problem areas.

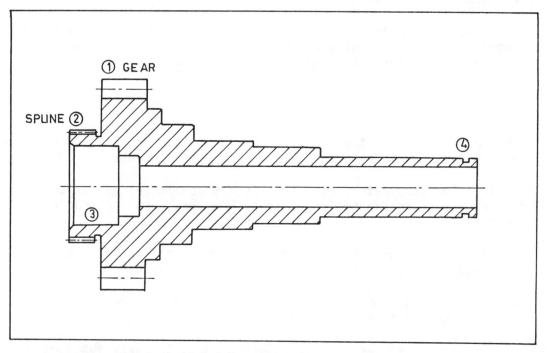

Figure 9-2. Countershaft design which includes typical heat treat problems.

Point 1 shows a coarse pitch gear in a heavy mass section. Requires long carburization time to reach the necessary case depth.

Point 2 shows a fine pitch spline with small thin teeth. Exposed to prolonged furnace time the spline teeth will through harden, resulting in brittleness and tooth breakage under load.

Point 3 illustrates an internal bearing bore. Will have substantially less case depth than the gear because it is not exposed freely to the carbon atmosphere. Subsequent grinding will reduce the case depth even further.

Point 4 shows thin wall and low mass section. Heats up and reaches carburization temperature faster than the gear resulting in longer carburization time and deeper case. Overcarburization will lead to through hardening and cracks.

A finished part as shown in Figure 9-2 with a general case depth specification of 0.030 to 0.050 is likely to end up having following actual case depths:

* Point 1: Gear tooth flanks: 0.03 - 0.05
* Point 2: Spline teeth: 0.04 - 0.06
* Point 3: Internal bore: 0.02 - 0.04
* Point 4: Stem   0.04 - 0.06

10
Grinding

OVERVIEW

Grinding operations machine critical features to final print tolerance. Examples of critical part features include internal and external bearing diameters, and length dimensions controlling assembly stack-ups and end play.

In general, grinding operations:

* Machine dimensions to precise tolerances,
* Create fine surface finish and texture, and
* Establish accurate form tolerances.

Dimensional Tolerances

Accurate tolerances are difficult to maintain in parts that are processed through heat treatment. Distortion, warp, and dimensional variation due to exposure to high temperature followed by shrinkage in the quench, affect the accuracy of hardened components. Heat-treated parts are therefore semi-finish machined prior to heat treatment, followed by finish grinding of critical dimensions after hardening.

Grinding operations are normally planned for diameter tolerances smaller than 0.004 in. (0.10 mm) and length dimensions which must be held within less than 0.004 in. (0.10 mm). Routinely achievable grinding tolerances are in the range of 0.0004 to 0.001 in. (0.010-0.025 mm) for diameters, and 0.001 to 0.002 in. (0.025-0.050 mm) for length dimensions. With special precautions, parts can be ground to even closer tolerances.

Non-heat-treated parts can be machined to final print specifications by precision turning and boring operations. Grinding of soft parts is recommended when extreme accurate tolerances of less than 0.001 in. (0.025 mm) are specified.

Surface Finish and Texture

Abrasive grinding processes are ideally suitable to produce fine surface finishes on hardened and soft parts. Metal cutting processes such as turning, boring, milling, broaching, etc., have a high stock removal rate but are limited by the roughness of the cut surface. Conversely, abrasive processes such as grinding, honing, polishing, lapping, and superfinishing, are characterized by low stock removal rate but achieve very fine surface quality. The lower limit for normal application of metal cutting processes is reached around 63 microinches (1.6 micrometers). Finer finishes can be produced with cutting operations, but the most economical method is found in abrasive technology. Table 10-1 provides an overview of achievable surface roughness with various machining methods.

Table 10-1
Surface Roughness Produced by Common Production Methods

Micrometers	12.5	6.3	3.2	1.6	0.8	0.4	0.2	0.1	0.05
Microinches	500	250	125	63	32	16	8	4	2
Sawing	XXXXX	XXXXX	ZZZZZ						
Planing		XXXXX	XXXXX	ZZZZ					
Drilling		XXXXX	XXXZZZ						
Milling			XXXXX	XXZZZ					
Broaching			XXXXX	XXXZZZ					
Reaming				XXXXXXZZZZ					
Boring		XXXXXXX	XXZZZZ	ZZZZ					
Turning		XXXXXXX	XXZZZZ	ZZZZ					
Burnishing					XXXXX	XXXXX	ZZZZZ		
Grinding				XXXXXX	XXXXX	XXXXX	ZZZZZ	ZZZZZ	
Honing					XXXXXX	XXXXX	XXXXX	ZZZZZ	ZZZ
Polishing					XXXXX	XXXXX	ZZZZZ	ZZZ	
Lapping					XXXXX	XXXX	XXXXX	XXXXX	ZZ
Superfinishing							XXXXX	XXXXX	ZZ

XXXXX : Roughness range with average applications
ZZZZZ : Roughness range with special applications

Lapping, superfinishing, and polishing are considered surface finishing processes for parts which are dimensionally finish machined. Their objective is to produce surface finish, lay and texture required for optimum performance of sliding components. The purpose of superfinishing is to remove tool marks, chatter marks, fragmented metal, etc., from the surface and to generate the texture required for service life, performance, and reliability of the product. The amount of metal removed with superfinishing is in the order of 0.0002 inch (0.005 mm). Surface finishes between 2 and 8 microinches (0.05-0.2 micrometers) can be obtained with superfinishing and lapping, and between 4 and 16 (0.1-0.4 micrometers) with polishing.

Roller burnishing is a non-metal removing process that improves surface finish of parts made of soft steel or cast iron. The smoothening effect is obtained by applying pressure with a hardened steel tool which rolls over the surface.

Cylindrical and surface grinding produce finishes in the range of 32 to 8 microinches (0.8-0.2 micrometers). Stock allowance for grinding can vary from 0.005 to 0.040 inch (0.127-1.00 mm). Gears and shafts used in automotive, truck, and agricultural drivelines are machined with grinding stock allowances of 0.010 to 0.015 in. on diameters (0.25-0.38 mm), and 0.005 to 0.010 in. (0.127-0.254 mm) on faces.

Grinding stock allowances are selected carefully to ensure that:

* minimum grinding time is required to reach final print dimensions.
* minimum amount of case depth is removed, to avoid loss of hardness and to maintain sufficient resistance to contact pressure.
* ground surfaces clean up completely.
* tolerance stack up conditions, including heat treatment distortions are fully eliminated.

Form Tolerances

Form tolerances and tolerances which control the relationship between features also are affected negatively by heat treatment.

Examples of form tolerances are flatness, straightness, roundness, and cylindricity. Parallelism, perpendicularity, and runout are examples of tolerances which control the relationship between form features. These tolerances can be machined to very accurate specifications in grinding operations.

A common example is a shoulder adjecent to an external bearing journal which must be held square to that journal within 0.001 in. (0.03 mm). To meet the squareness tolerance, the shoulder is ground simultaneosly with the bearing diameter on an angle head grinder with a wheel which is precisely formed at a 90 degree angle. Perpendicularity of shoulder and diameter is guaranteed by the intrinsic accuracy of the process. Since shoulder and bearing journal are ground in the same plunge cut, accuracy is achieved at no extra cost.

Similarly, concentricity between diameters is easily obtained by grinding these diameters in one setup using a specially profiled grinding wheel or with multiple wheels mounted on the same arbor. Other geometrical tolerances for which grinding is recommended are: roundness of bearing journals, straightness of bearing journals and bushing diameters, flatness of surfaces, parallelism and runout of surfaces and diameters.

GRINDING METHODS

Three basic types of grinding machines are used in production operations.

* External grinding machines that grind diameters, tapers, or contours located at the outside of rotational parts.
* Internal grinding machines that grind diameters, tapers, or contours located at the inside of rotational parts.
* Surface grinding machines that grind flat surfaces on rotational or prismatic parts.

External grinders are built in two configurations. They are the angle head type and the plain cylindrical type. On angle head grinders, the grinding wheel is inclined at an angle of 50 to 60 degrees relative to the axis of the workpiece. The direction of infeed is also angular. This results in radial and axial stock removal.

On plain cylindrical grinders, the grinding wheel is perpendicular to the axis of the workpiece. The direction of infeed also is perpendicular to the axis of the workpiece resulting in radial stock removal only.

Shafts are typical parts machined on external grinders. They are located between centers and driven by an attachment in the spindle face plate. Smaller parts are mounted on expanding collet arbors which are loaded between the centers of the machine.

Internal grinders have an oscillating grinding spindle which can be set parallel or at an angle with the axis of the workpiece. The direction of infeed is perpendicular to the ground surface. Because of the slender design of internal grinding spindles, the infeed value per oscillation is very small to avoid deflection. Internal grinders often are equipped with an additional spindle to grind faces simultaneously with bores. Parts are clamped and located in precision diaphragm chucks that assure concentricity of the ground bore with the locating diameter. Gears are centered and clamped in the pitch diameter with three equally spaced fingers that locate in tooth spaces of the gear.

Surface grinders can either be built with horizontal spindle, or with vertical spindle. In the horizontal spindle type, grinding occurs with the periphery of the grinding wheel. Vertical spindle machines grind with the face of a solid or segmented grinding wheel. Surface grinders are normally equipped with a magnetic table to hold the parts during the operation. Machine tables for surface grinding machines are built in the rotary table version, or with a reciprocating table. Rotary tables are very convenient for mass production as they can be loaded and unloaded without interruption during the machine cycle.

EXAMPLES OF CYLINDRICAL GRINDING OPERATIONS

Plunge Grind

For a plunge grinding cycle, the grinding wheel must overlap the width of the diameter to be ground. The diameter is finished in one plunge cut of the wheel (Figure 10-1). Different diameters, tapers, and even contours, can be ground with this method provided there are no shoulders perpendicular to the work axis which need to be ground. Most cylindrical grinders have a maximum wheel width of 4 in. Specially designed machines go up to 6 in. wheel width.

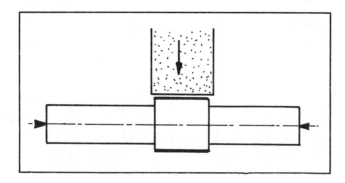

Figure 10-1. Plunge grinding of one diameter.

Traverse Grind

Traverse grinding is a grinding method used when the diameter to be ground is much longer than the maximum permissible grinding wheel width. The table, with the workpiece located between centers, reciprocates in front of the grinding wheel. Wheel infeed occurs between table reversals. The diameter is finished in several traversing passes of the wheel, with the last few passes without infeed for spark-out (Figure 10-2).

Figure 10-2. Traverse grinding of a shaft diameter when the grinding length exceeds the wheel width.

Multiwheel Grind

This form of grinding is used to grind two or more diameters spaced far apart in a single plunge of the wheelhead. The operation is completed in one setup by installing multiple grinding wheels on an arbor. A multiwheel grinder is usually dedicated to one specific operation. Change-over is rather cumbersome. Multiwheel grinding is typically used to grind the main bearing journals of crankshafts and camshafts in one setup (Figure 10-3).

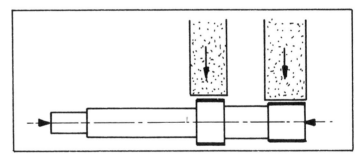

Figure 10-3. Grinding of two diameters with a pair of wheels mounted on the same arbor.

CNC Grind

CNC cylindrical grinding offers an economic and flexible alternative to multiwheel grinding. Multiple diameters are ground with a single grinding wheel with CNC-controlled axis travel. The operation is completed in consecutive plunge cuts. CNC grinding provides flexibility to handle many different parts with a universal machine setup. CNC grinding also offers the opportunity to reduce labor by combining two or three plunge grinding operations in one setup (Figure 10-4).

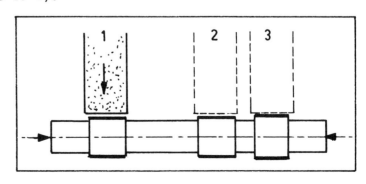

Figure 10-4. CNC grinding of three diameters with a single wheel.

Groove Grind

A groove grinder is a special version of a cylindrical grinding machine. It is used to finish grind one, two, or more grooves in a single plunge cut of the grinding wheel. The grinding wheel is formed to the right profile with a diamond dressing roll. Grooves can be finish ground from the solid material or may be precut on a lathe. Grinding of grooves may be required to hold surface finish specifications in the groove, or to maintain close length tolerances between groove and other features. Typical applications are grinding of snap ring grooves through splines, and grinding of seal grooves in shafts (Figure 10-5).

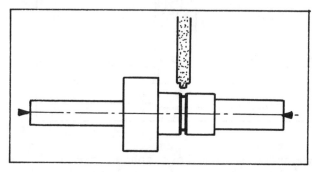

Figure 10-5. Groove grinding with a form wheel.

Centerless Grind

Centerless grinding is a widely applied grinding method to finish external diameters of pins, spool valves, rails, shafts, bushings, etc. It is capable of producing parts at rates of several hundred pieces per hour, while maintaining precise tolerances and fine finishes. Diameter tolerances of 0.0002 in. (0.005 mm) and surface finish of 10 microinches are consistently held in production. Grinding is continuous with the work being fed between the grinding and regulating wheels and ejected at the opposite side (Figure 10-6). Multiple diameter and contoured parts can also be centerless ground by the infeed method (Figure 10-7).

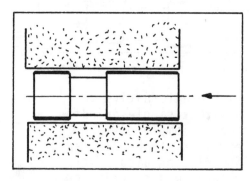

Figure 10-6. Centerless grind with the through-feed method.

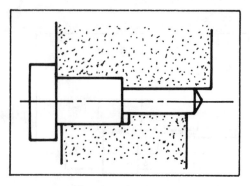

Figure 10-7. Centerless grind with the in-feed method.

ANGLE HEAD GRINDING

Diameter and Face Grind

A diameter and adjecent face are ground simultaneously in one plunge cut of the grinding wheel. To allow this cycle, the right angle contour of the wheel must overlap both the diameter and the face width. The shoulder is held square to the diameter by the intrinsic accuracy of wheel dressing (Figure 10-8).

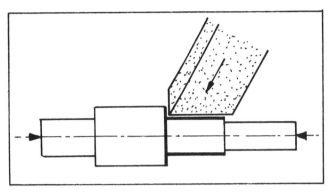

Figure 10-8. Grinding of diameter and face on an angle head grinder.

Multiple Diameters and Faces

More than one diameter and shoulder are ground in one setup with a specially profiled grinding wheel. This process can be used with wheel widths up to 6 in. Length dimensions between ground faces are held very accurately. The grinding wheel is formed with a precision template and single point diamond dresser, or with a diamond dressing roll in high-volume applications (Figure 10-9).

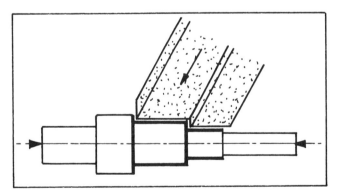

Figure 10-9. Grinding of two diameters and two faces on an angle head grinder.

CNC Angle Head Grind

Multiple diameters and shoulders are ground in one setup with a single wheel that travels to each position and finishes each diameter in one plunge cut. CNC angle head grinding can handle many different parts with universal tooling and is ideal for batch production. Besides saving labor by combining several setups in one operation, CNC grinding also is more economical in tooling because it does not require expensive templates and diamond dressing tools (Figure 10-10).

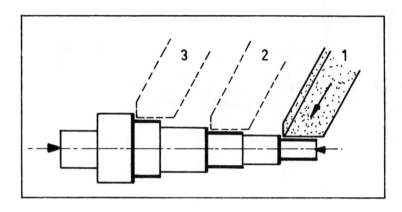

Figure 10-10. Grinding of three diameters and three faces on a CNC angle head grinder with a single wheel.

INTERNAL GRINDING

Bore Grind

The bore is ground with a small diameter grinding wheel mounted on an oscillating spindle. To avoid deflection of the spindle, the infeed value is kept very small. At the end of the cycle, the spindle oscillates several times without infeed to eliminate taper and out of round. The spindle can be set at an angle to grind internal tapers (Figure 10-11).

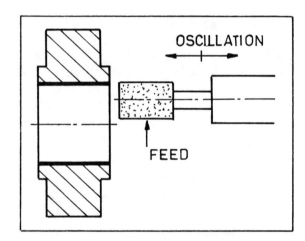

Figure 10-11. Bore grinding.

Bore and Face Grind

The bore and one thrust face are ground simultaneously with two independent grinding spindles. The face is ground with a cup wheel with infeed perpendicular to the face. Squareness between face and bore is maintained by the alignment of both spindles. The cup wheel produces a cross hatched pattern on the thrust face (Figure 10-12).

Bore and Face Grind with Angle Head Wheel

When an angle head wheel grinds the thrust face, line contact occurs between wheel and workpiece resulting in a circular pattern on the face. Circular grind may be required to improve wear conditions on rotating components (Figure 10-13).

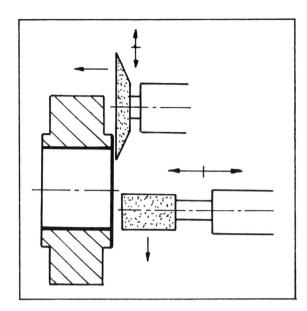

Figure 10-12. Bore and face grind with two independent grinding spindles.

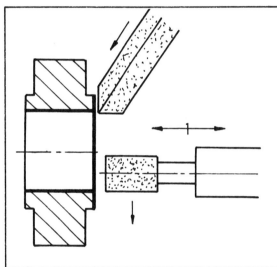

Figure 10-13. Bore and face grind with two independent grinding spindles. The face is ground with an angle head wheel.

Bore and Face Grind with One Spindle

Grinding of bore and face in the same cycle also is possible on grinding machines with only one grinding spindle. Both grinding wheels are then mounted on the same spindle as shown in Figure 10-14.

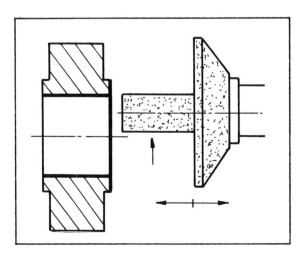

Figure 10-14. Bore and face grinding wheels are mounted on the same spindle.

Grinding of a Recessed Face

When the thrust face is recessed deep into the part, the methods shown in Figure 10-12, Figure 10-13, and Figure 10-14 are not possible. The face can be ground with the front of the internal grinding wheel. This method creates a cross-hatched pattern on the surface (Figure 10-15).

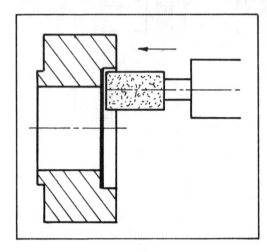

Figure 10-15. Bump grinding of a recessed face.

Grinding of a Bore and a Recessed Face with Circular Lay

A circular lay can be obtained in a recessed face by inclining the spindle relative to the work axis. The conical-shaped grinding wheel creates line contact between face and grinding wheel. This results in circular grind pattern (Figure 10-16).

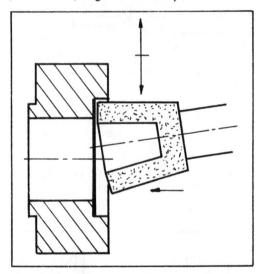

Figure 10-16. Bump grinding with a conical-shaped wheel produces a circular lay on the face.

CNC Internal Grind

Several diameters and faces can be ground in the same setup on a CNC-controlled internal grinding machine. The CNC grinding method gives good quality results with respect to relationship between geometrical features. Diameters are ground concentric to each other. Faces are held square to the diameters and parallel to each other (Figure 10-17).

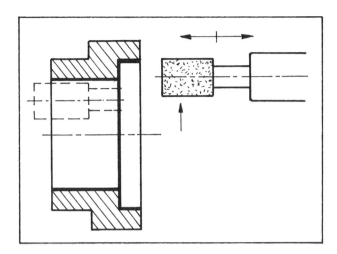

Figure 10-17. CNC grinding of two internal diameters and one shoulder.

SURFACE GRIND

Peripheral Grind

On this type of grinding machine, the spindle is horizontal and the workpiece is ground with the periphery of the grinding wheel. The parts are clamped on the magnetic machine table. Part configuration must be such that no interference occurs with the grinding wheel. The machine table is either a rotary type, or a reciprocating type. This type of surface grinder is often found in toolrooms (Figure 10-18).

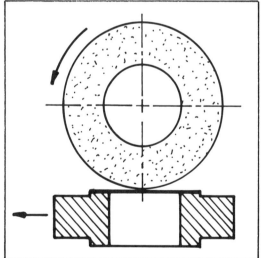

Figure 10-18. Surface grinding with the periphery of the wheel.

Vertical Disc Grind

Vertical disc grinders are frequently used in high-volume production to finish grind flat surfaces while maintaining width tolerances of 0.001 in. (0.03 mm). The parts are loaded on a large rotary table with magnetic clamping action. Load and unload is continuous, without stopping the rotation of the work table. It is essential for this type of operation that the part has no protrusions beyond the face to be ground (Figure 10-19).

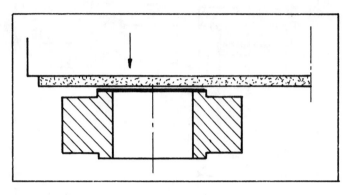

Figure 10-19. Surface grinding with the face of a solid or segmented wheel.

Horizontal Disc Grind

Horizontal disc grinders grind the end faces of shafts or they grind the side faces of rings and disc type parts. Machines are configured with stationary grinding spindles, and with fixtures that feed the parts through the grinding discs in a continuous motion (Figure 10-20).

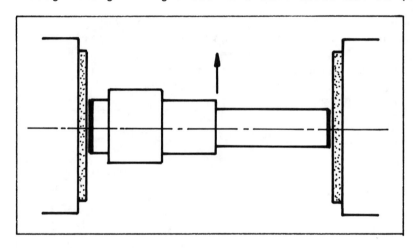

Figure 10-20. Surface grinding of the end faces of a shaft between two grinding discs.

11
Process Planning Examples

The following examples are taken from actual production. They illustrate the various methods of gear and spline manufacturing explained in the previous chapters. They also illustrate how these operations fit in the overall process planning sequence.

Example 1: Idler Gear

In assembly, a needle bearing is installed in the bore and the gear is mounted on a shaft. Functionally, the gear must run concentric with the bore. In addition, the hub faces must be square to the bore and parallel to each other. (Figure 11-1).

To achieve concentricity between the bore and the external features of the gear, the bore is machined to close tolerance in the blanking operation 30. The bore is then used as a radial locator in hobbing and shaving (Figure 11-2).

For accurate location in hobbing and shaving, the bore diameter is held within a tolerance of 0.001 in. (0.025 mm). The rimfaces are turned square to the bore. Operation 30 is a finish turning setup which finishes the bore diameter within 0.001. It maintains the squareness of one rimface relative to the bore by turning that face in the same setup. During the operation, the gear is located against the opposite rimface. This location guarantees parallelism of both faces.

After heat treatment, in operation 90 the gear is held in a pitch line diaphragm chuck (which locates with fingers in three tooth spaces). The bore is ground to part print tolerance. Locating in the pitch diameter of the gear ensures concentricty between the ground bore and the gear pitch diameter.

The squareness tolerance of both hub faces relative to the bore is maintained by grinding one face in the same setup as the bore in operation 90. Grinding the opposite hub face in operation 100 with the gear mounted on an expanding arbor which locates in the ground bore accomplishes the squareness requirement for the other face.

Because of the accuracy of CNC lathes, a separate finish turning operation, as shown in this example, is not always required. A good quality blank also can be produced in just two turning operations. The bore is then rough turned in operation 10, and finish turned in operation 20.

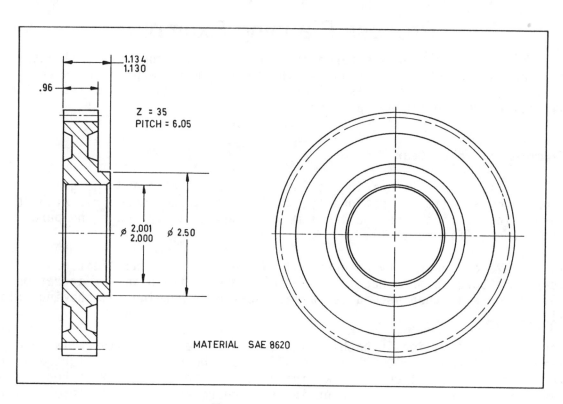

Figure 11-1. Idler gear.

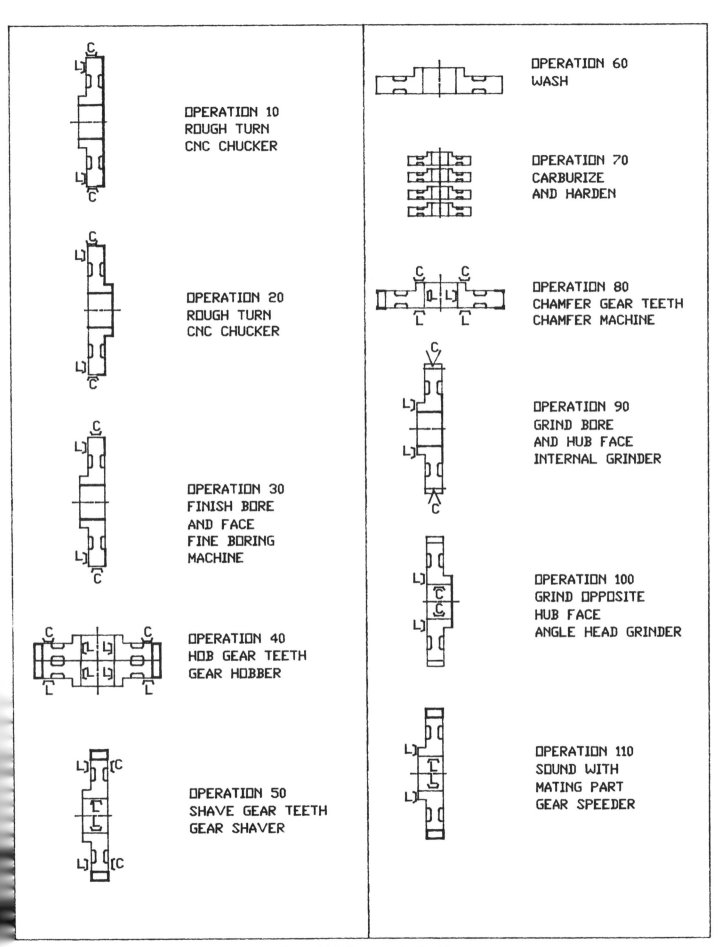

Figure 11-2. Process plan for idler gear.

Example 2: Cluster Gear

The difficulty in the manufacturing process for the cluster gear shown in Figure 11-3 consists in holding concentricity between internal spline and the two external gears. In assembly, the gear is mounted on a shaft and located in the internal spline. For trouble-free assembly, a minimum clearance is required between mating splines. Too much clearance will, however, result in vibrations and noise. Maintaining spline tolerances is, in this case, a critical requirement. The requirement is made particularly difficult by the variability of the heat treatment process.

After rough turning in operation 10, the internal spline is broached with a concentricity broach. This broach cuts the minor diameter together with the spline teeth. The gear blank is then pressed on a solid splined arbor. The outside diameters are finish turned, concentric with the spline, and the rim faces are cut square to the spline in operation 30 (Figure 11-4).

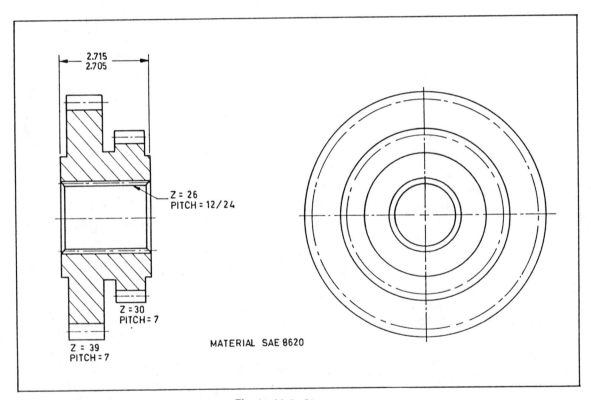

Figure 11-3. Cluster gear.

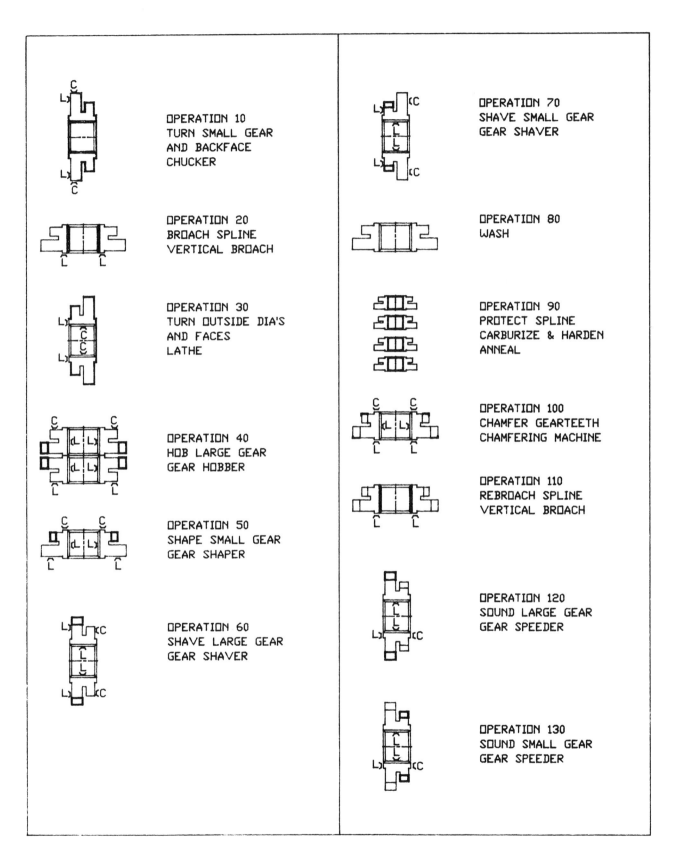

Figure 11-4. Process plan for a cluster gear.

The gear is located on the minor diameter with solid arbors in the following operations. The large gear is hobbed in operation 40 (Figure 11-4), and the shoulder gear is shaper cut in operation 50 (Figure 11-4). Both gears are finish shaved.

Before being loaded in the furnace, the internal spline is coated with a carburization inhibitor. This coating prevents hardening and maintains adequate machinability of the spline, after heat treatment.

In operation 110 (Figure 11-4), the internal spline is finish broached to final print tolerance. The pilot of the rebroach is equipped with a locator. This locator aligns the spline teeth with the cutting teeth of the broach.

This process guarantees accurate and consistent spline tolerances. Hardness of the spline must, however, be controlled carefully to avoid tool breakage or premature broach wear.

Example 3: Cluster Gear

This cluster gear has two external gears and a synchronizer spline (Figure 11-5). Gears and spline must run concentric with the axis of two internal and opposite needle bearing bores.

The rough material is a forging with a pierced thru central hole. Blanking occurs in two turning operations on CNC chuckers. Because of the length-to-diameter ratio, the part is hobbed, shaved, and shaped, while located between centers.

After heat treatment, the needle bearing bores must be ground concentric with the axis of the two gear pitch diameters.

The problem of maintaining critical runout tolerances between gears and internal bearing bores is solved by combining the internal grinding operations in one setup. This setup is on a special double-end internal grinder with centerdrive chuck.

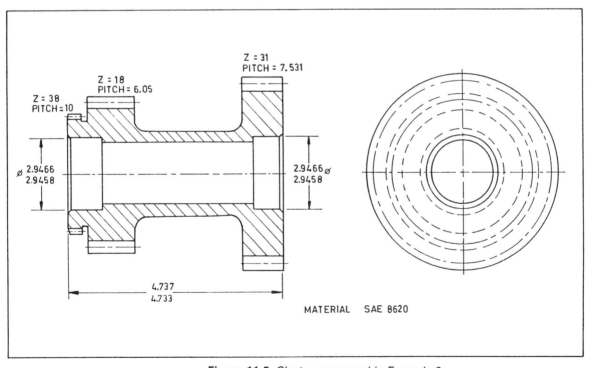

Figure 11-5. Cluster gear used in Example 3.

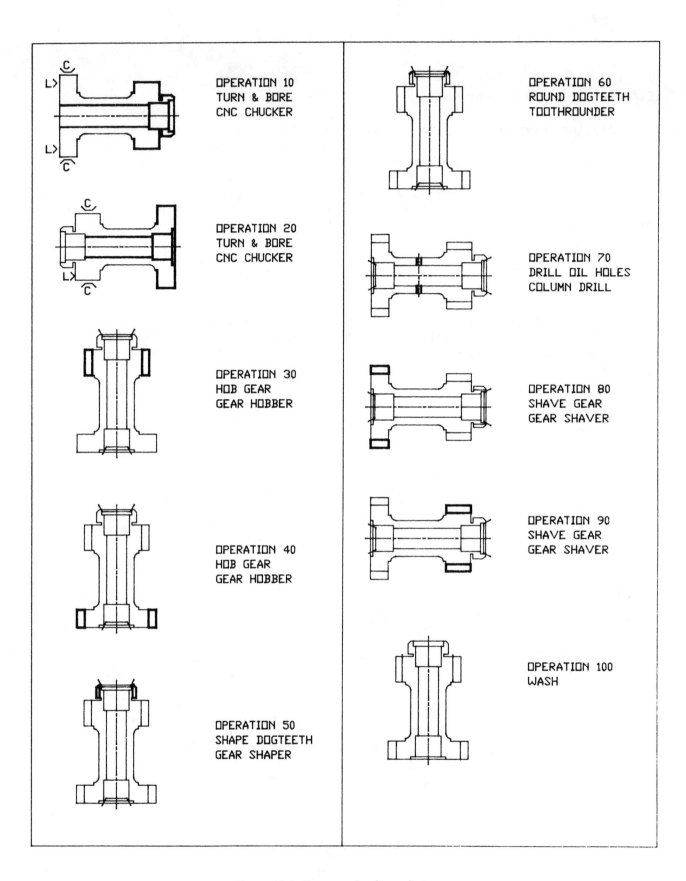

Figure 11-6. Process plan for a cluster gear.

In operation 140 (Figure 11-7), the part is located and clamped in a double diaphragm chuck with pitch line locators. These locators center and equalize the runout between both gears. The bores are then finish ground with opposite grinding spindles. The end faces are finish ground in the same setup.

The alignment of both bores is guaranteed by the alignment of the grinding spindles. Runout between bores and gears is controlled by the clamping method. With this method of grinding, the bore diameters are held within 0.0008 in. (0.02 mm) tolerance and concentric with each other within 0.001 in. (0.03 mm) total indicator reading. The length dimension between the end faces is held within 0.002 in. (0.05 mm) tolerance. Squareness of end faces relative to bores is held within 0.001 in. (0.03 mm).

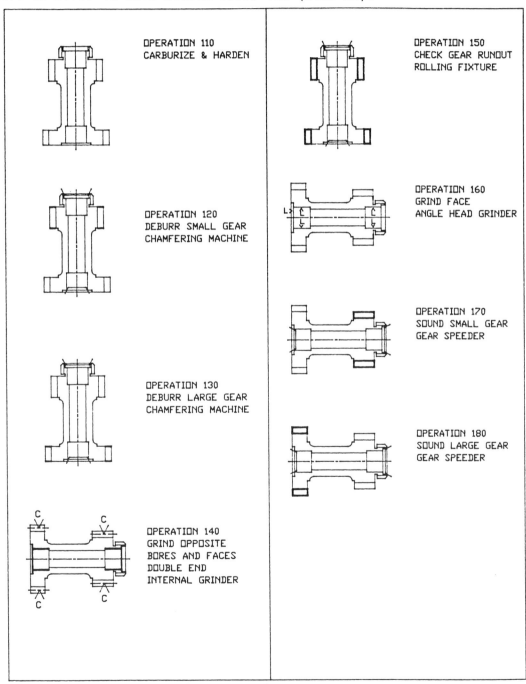

Figure 11-7. Continuation of the process plan for a cluster gear.

Example 4: Countershaft

A countershaft design which incorporates three or more gears and splines is very common in transmission design (Figure 11-8). This countershaft has three external gears. In assembly, these gears must run concentric with the axis of the bearing journals at each end of the shaft.

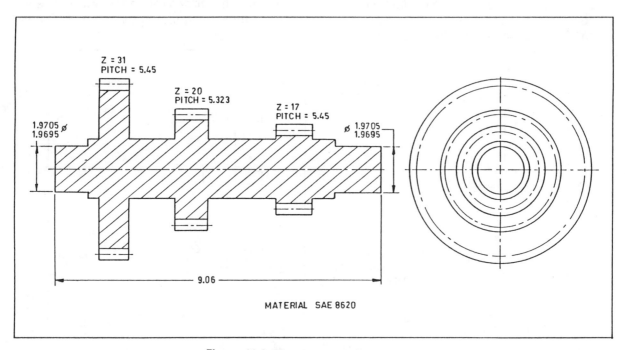

Figure 11-8. *The countershaft used in Example 4.*

Centers are made in operation 10 (Figure 11-9) and are used for location in all subsequent operations. Centers are not functional within the part. However, they provide an accurate and consistent method of locating parts. They are very important in achieving concentricity between features machined in different operations. For this reason, the process plan includes center lapping operation 120 to clean up the centers after heat treatment.

Two of the gears are hobbed and shaved, while the middle gear is shaped and shaved. Insufficient clearance for hob runout between the middle gear and the 31-tooth gear makes shaping necessary.

One hobbing operation could be eliminated with the introduction of a CNC hobbing machine in the line. The 17-tooth gear and the 31-tooth gear have the same diametral pitch and pressure angle. Both gears can be hobbed in one cycle on a CNC hobber.

Although not included in the processing, a straightening operation after carburizing and hardening may be necessary. This is true when heat treatment distortion is excessive.

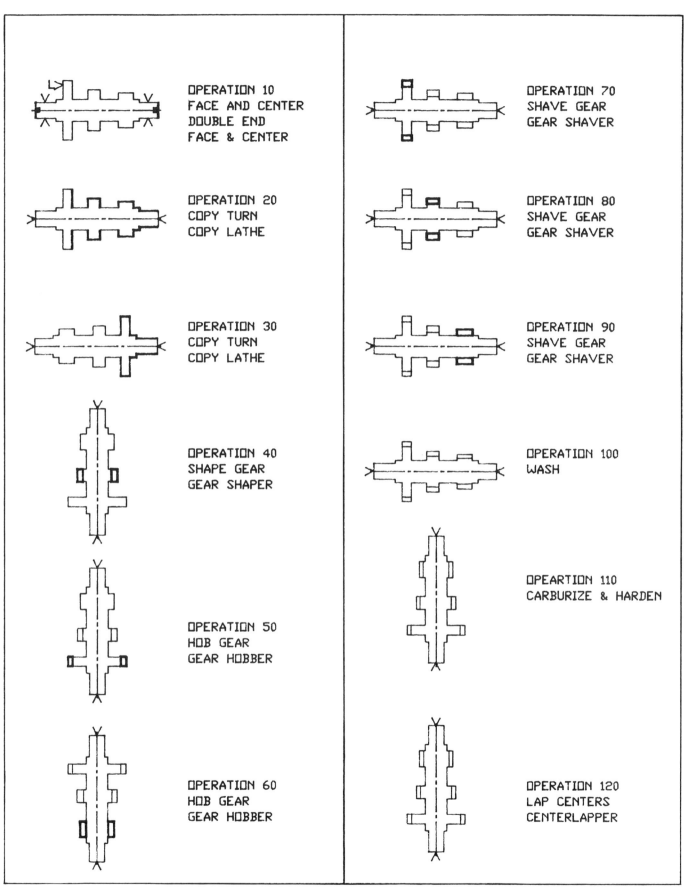

Figure 11-9. A process plan for the countershaft.

The gears are chamfered after heat treatment to eliminate nicks, burrs, and other tooth damages. This type of damage may occur during handling and transport.

This process plan includes a hard finishing operation of the gears. To improve quality, each gear is honed in operations 180, 190, and 200 after heat treatment.

As a final inspection operation, the runout of the gears is checked in a rolling fixture. The shaft is mounted between centers in a rolling fixture and each gear is rolled against a master gear. Total gear runout is measured by charting the radial displacement of the centers during rolling.

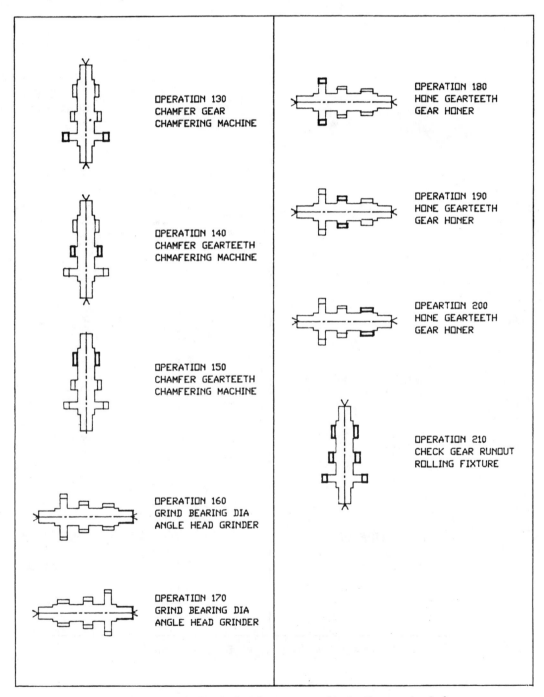

Figure 11-10. Continuing the process plan for the countershaft.

Example 5: Internal Planetary Ring Gear

This example (Figure 11-11) is an internal planetary ring gear with adjacent shoulder which must be shaper cut and shaved. Only the gear teeth are induction hardened. The rest of the part is allowed to remain soft. Heat treatment is a potential source of quality problems because the part is asymetric and susceptible to heat treatment distortions.

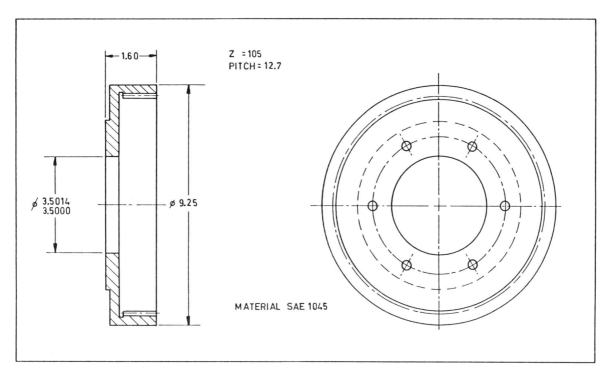

Figure 11-11. Internal planetary ring gear.

To keep induction hardening distortions to a minimum, the gear blank is turned in two phases with an intermediate stress relieving operation.

Starting with a forging, the part is rough-turned in operations 10 and 20 with heavy chip removal cuts.

A stress relieving operation between rough and finish turning eliminates residual forging stresses. The operation also eliminates added machining stresses induced by the heavy metal removal in operations 10 and 20 (Figure 11-12).

In the finish turning operations, the gear is clamped with low pressure in chucks with enveloping jaws to minimize distortion. Light cuts are taken to finish the part contour.

The objective of the intermediate stress relief is to minimize and control the variation of induction hardening distortions. This particular gear configuration, an internal shoulder gear, does not permit hard finishing.

Heat treatment development is critical for this part. All hardening distortions must be compensated by modifications in the shaping and shaving operation. Possible corrections are: taper shaping and shaving of the gear teeth to compensate taper conditions resulting from hardening, and crown shaving to prevent end loading of the gear teeth.

After hardening, the location bore is ground on an internal grinding machine with the part located and clamped in the pitch diameter of the gear. This type of location ensures runout tolerances between gear and pilot diameter, to fall within print specification.

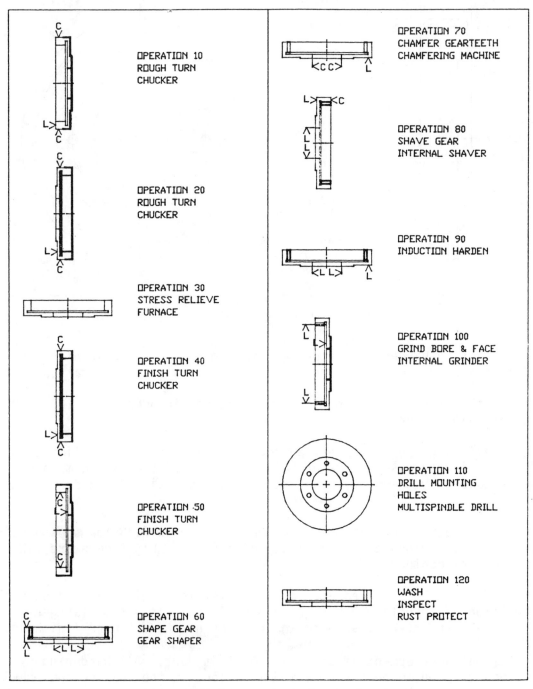

Figure 11-12. Process plan for the planetary ring gear.

Example 6: Transmission Countershaft

Typical for many driveline parts, the design of this transmission countershaft contains several different tooth forms (Figure 11-13). Each tooth form is machined with a special process. A 17-tooth gear is hobbed and shaved, a 28-tooth synchronizer spline is shaped, and a 32-tooth spline with 45 degree pressure angle is rolled from a solid blank diameter (Figure 11-14).

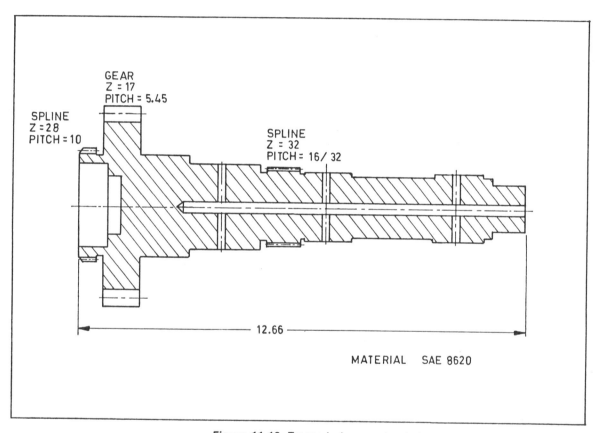

Figure 11-13. Transmission countershaft.

The 28-tooth spline and the 17-tooth gear are cut simultaneously in operation 50 (Figure 11-14). This cut takes place on a combination shaper-hobber, called "shobber," a method patented by Pfauter. The 28-tooth spline can be cut with or without backtaper on the shobber.

In the hobbing and shaping operation, the part is located on the outside diameter near the gear. To ensure concentricity of gear and spline with the other part features, the location diameter is ground in operation 40 (Figure 11-14). A diameter tolerance of 0.001 is held in the grinding operation.

The deephole drilling and crosshole drilling operations are usually performed after the gear cutting is finished. The deephole drill will eliminate or damage the location center in the end of the shaft. It may be necessary to remachine the center after the operation.

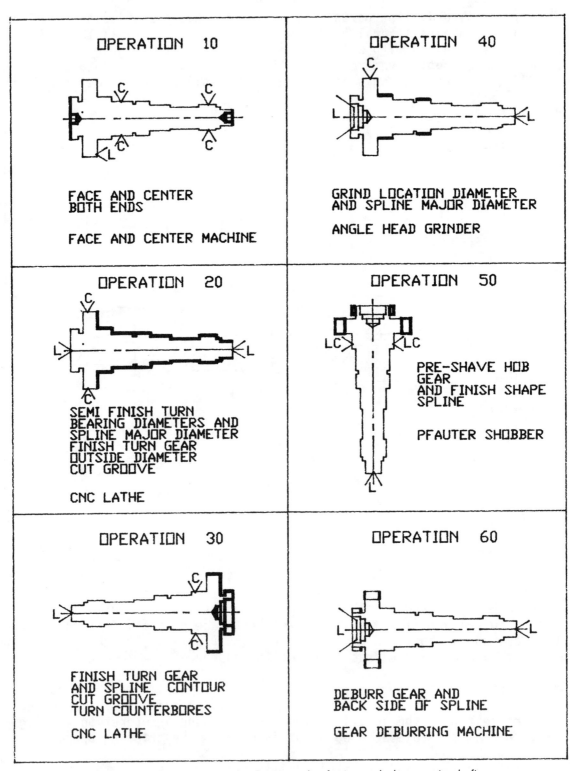

Figure 11-14. Process plan for transmission countershaft.

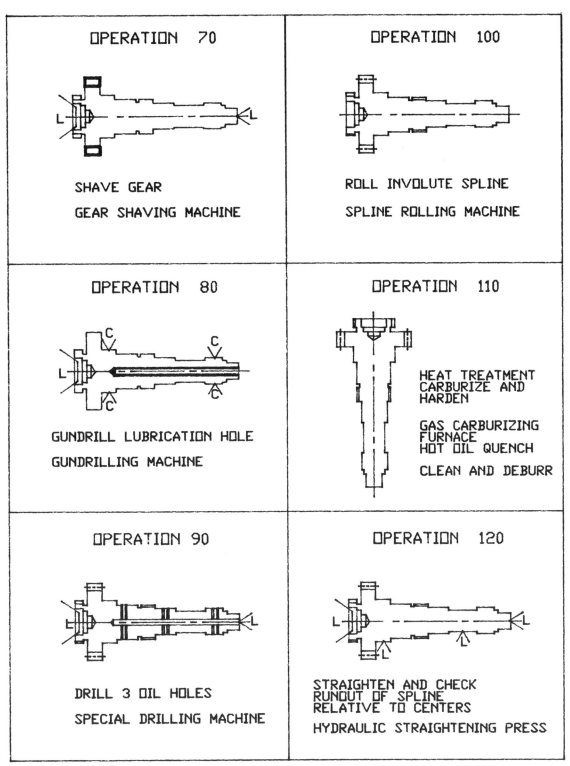

Figure 11-15. Continuation of the process plan for the transmission countershaft.

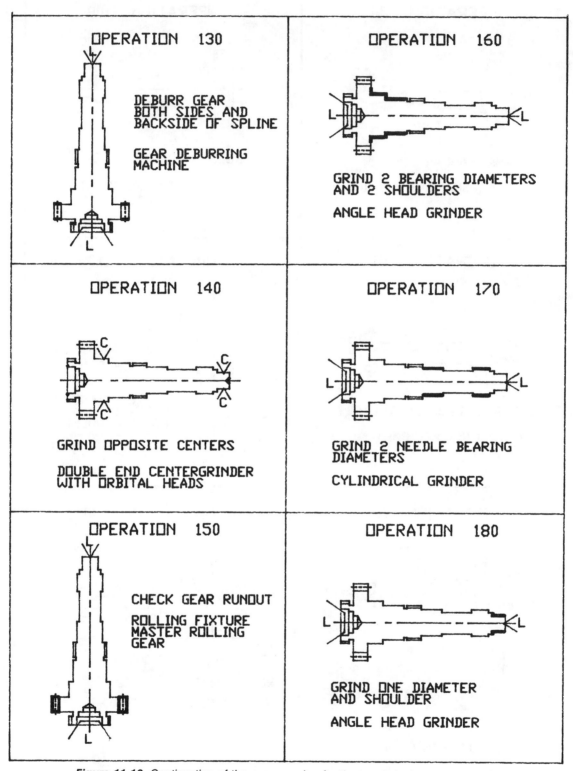

Figure 11-16. Continuation of the process plan for the transmission countershaft.

Because of the large number of grinding operations after heat treatment, all with very accurate tolerance specifications, the location centers are ground in operation 140 (Figure 11-16) on a double end center grinding machine.

Center grinding is an operation which is often performed as repair when parts cannot be brought within normal runout with straightening. In the center grinding operation, the part is located in the spline or gear. The position of the center is brought in line with the critical features.

Center grinding also will improve diameter grinding tolerances. This improvement is caused by reducing or eliminating roundness errors resulting from part wobble between centers.

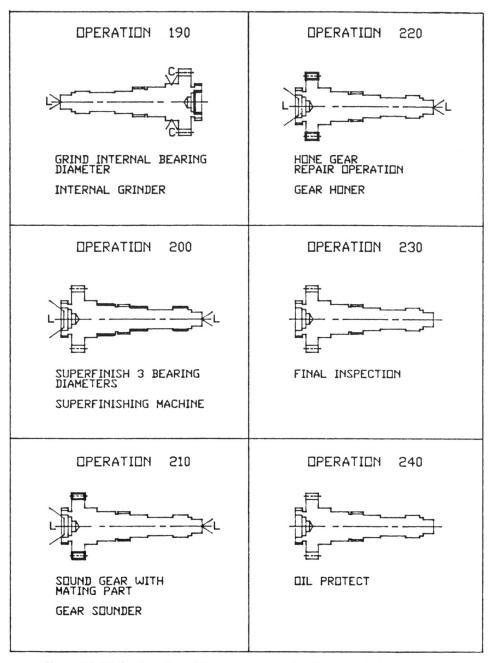

Figure 11-17. Continuation of the process plan for the transmission countershaft.

Example 7: Gear for Synchronized Transmission

This type of gear incorporates a synchronizer spline (Figure 11-18). This gear configuration is frequently used in manual shift transmission design. The gear is engaged by sliding a coupler over the spline. Once engaged by the coupler, a solid connection is created between gear and shaft.

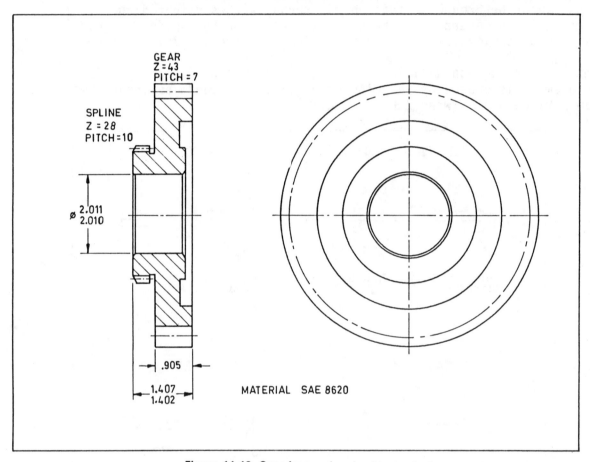

Figure 11-18. Gear for synchronized transmission.

The processing differs from Example 1 only in the machining operations for the 28-tooth spline for synchromesh engagement. The gear is hobbed and shaved, and the synchronizer spline is shaper cut.

Two specific features of synchromesh gears are shown in Figure 11-20. The first one is the tapered spline tooth which is form rolled in operation 80. Once the synchronizer coupler is engaged with the spline, the backtaper will lock the mating teeth together and prevent disengagement. Backtaper is cut on a shaper with tilting table, or on a shaper with tilted cutter spindle axis. Heavy backtaper requires an additional rolling operation on a special machine. Back taper on splines is applied to prevent jumping out of gear.

The second feature is the chamfering or tooth pointing of the spline teeth in operation 90. This operation is performed on a special tooth chamfering machine. Pointing of the spline teeth is necessary to facilitate the

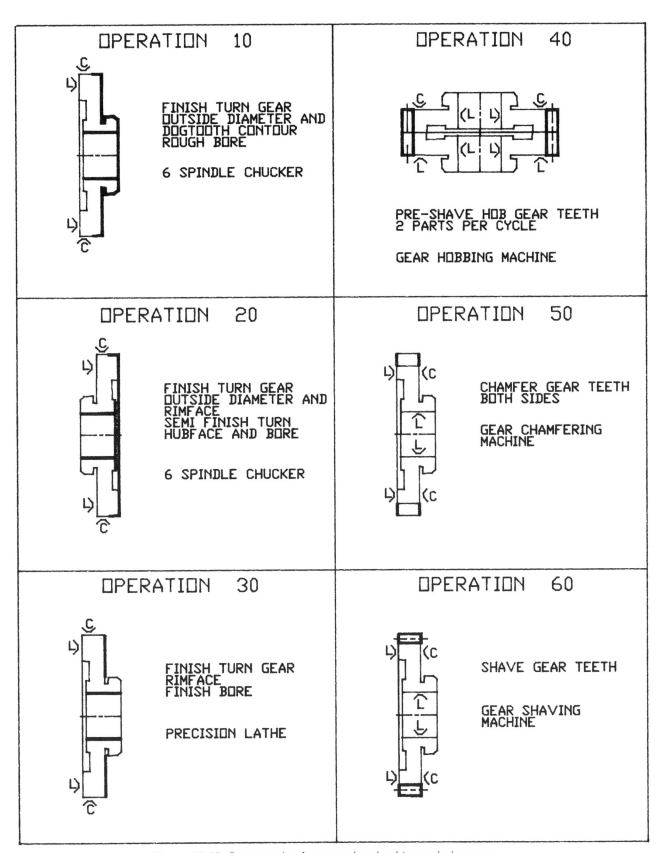

Figure 11-19. Process plan for a synchronized transmission gear.

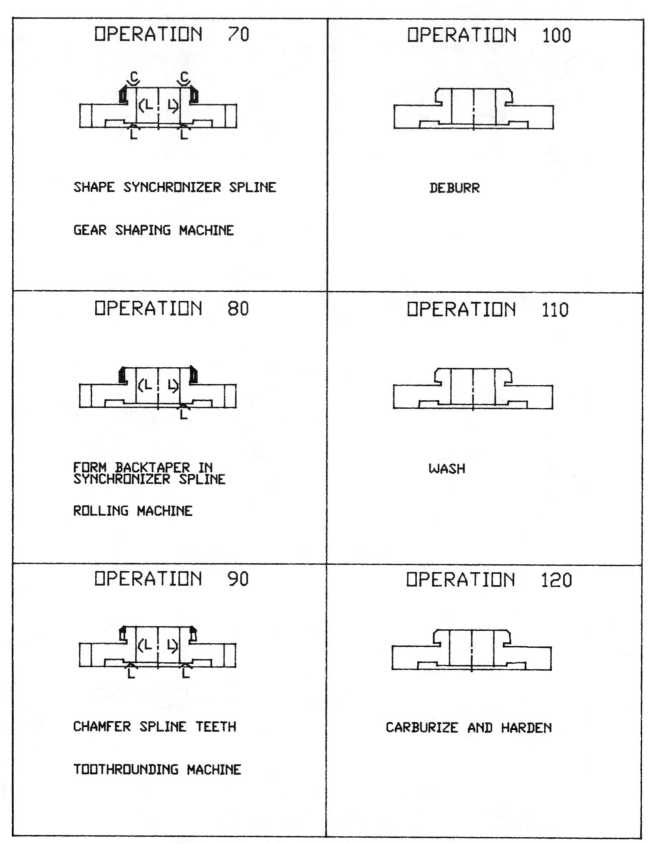

Figure 11-20. Continuation of the process plan for a synchronized transmission gear.

engagement of the sliding coupler with the synchronizer spline when both rotate at different speeds. Without chamfering or rounding the mating teeth would clash when forced together.

The spline is shaper cut in operation 70, back-tapered in a rolling machine in operation 80 (Figure 11-20), and chamfered in operation 90 (Figure 11-20). The rest of the process planning sequence is similar to Example 1.

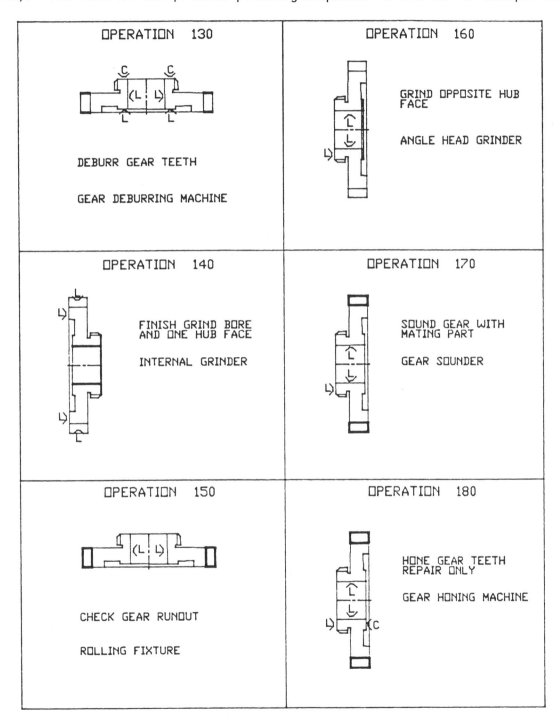

Figure 11-21. Continuation of the process plan for a synchronized transmission gear.

Example 8: Internal Planetary Ring Gear

The ring gear is used in the final planetary reduction of a tractor rear axle (Figure 11-22). This gear is produced in quantities of about 120,000 pieces per year.

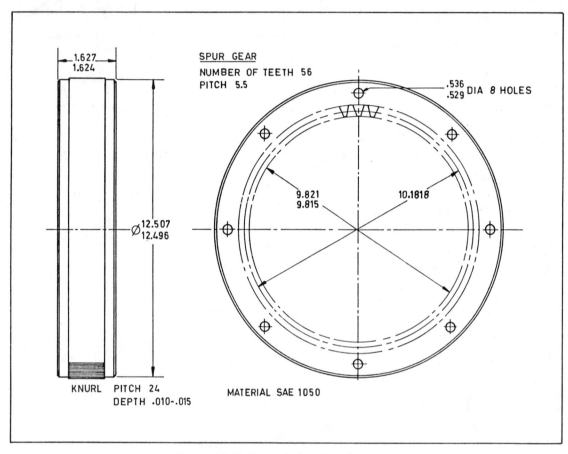

Figure 11-22. Internal planetary ring gear.

The part is made from a forged ring which is turned completely in three operations on single spindle chuckers. Three turning operations are required to maintain roundness tolerance. The heavy clamping pressure used in operation 10 (Figure 11-23) distorts the ring while it is clamped in the chuck. Upon release of the clamping pressure, the ring springs-back to free condition resulting in out-of-round of the turned diameter. In operation 20 and 30 (Figure 11-23), the ring can be clamped with reduced pressure on the already-turned diameters. Enveloping jaws cause minimum distortion. An accurately finished internal diameter is necessary to pilot the broach in operation 40 (Figure 11-23).

The internal spur gear is finish broached in a single pass with a full form finishing broach.

The outside diameter of the gear is knurled to provide a press fit in assembly with the cast iron rear axle housing.

The gear teeth are induction hardened to below the root diameter. The material section where the holes are drilled remains soft. Drilling and

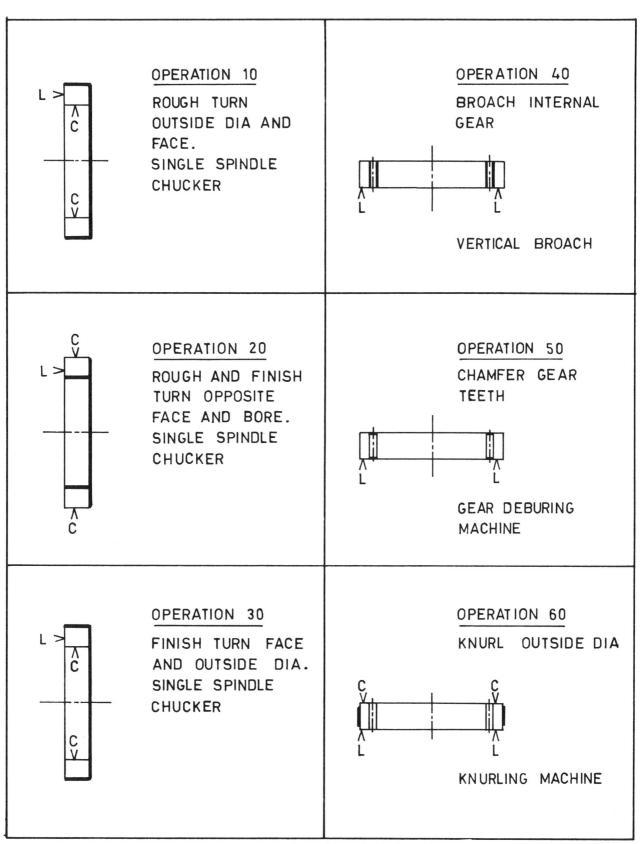

Figure 11-23. Process plan for an internal planetary ring gear.

reaming of the mounting holes is performed after heat treatment for two reasons. First, the position tolerance of the holes is not affected by heat treatment if the holes are drilled after induction hardening. Secondly, heat treatment distortion of the ring gear is not affected by the presence of holes which weaken the material section. With the holes, induction hardening would cause an 8-lobed out-of-round shape.

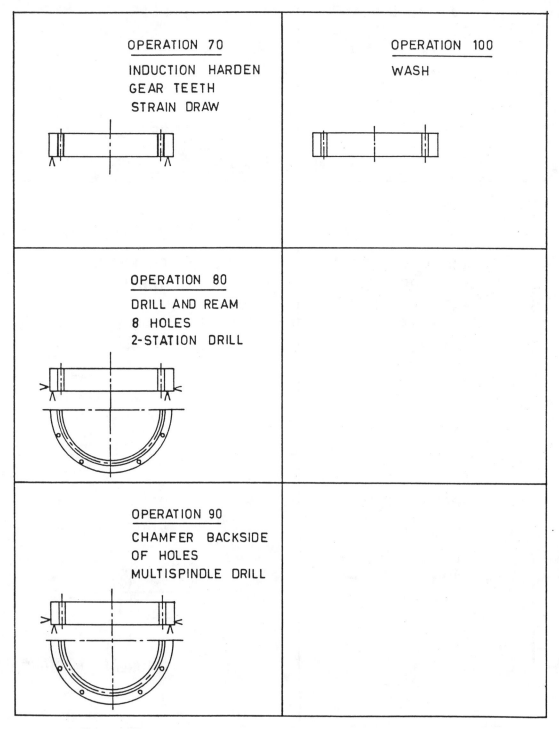

Figure 11-24. Continuation of the process plan for an internal planetary ring gear.

Example 9: Transmission Connector

The connector is made from SAE 1141 steel. Only the external spline teeth need to be induction hardened (Figure 11-25).

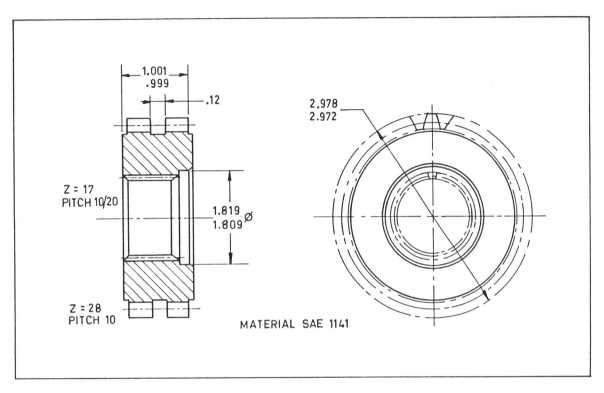

Figure 11-25. Transmission connector.

Because of the small outside diameter, the part is made most economically from barstock material. Rough turning is performed in one operation on a multispindle barchucker. In operation 10 (Figure 11-26), the bore is held within 0.004 in. (0.10 mm) tolerance to provide a good pilot diameter for the broach.

The minor diameter of the spline is cut concentric with the spline pitch with a concentricity broach.

After spline broaching, the part is loaded on a splined arbor and finish turned in operation 30 (Figure 11-26) on an automatic lathe. This operation eliminates perpendicularity errors. The step also eliminates drift which may occur with dulling broaches. The outside diameter is turned concentric with the spline. The side faces are cut square to the spline.

In operation 40 to 90 (Figure 11-26 and 11-27), the part is located on solid arbors in the minor diameter of the spline. The spline is hobbed, deburred,

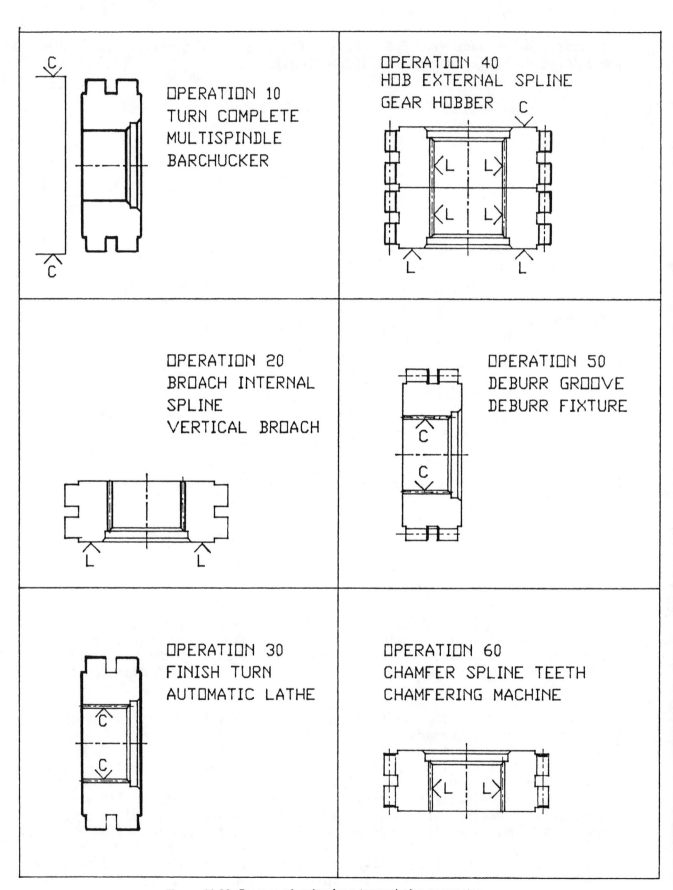

Figure 11-26. *Process planning for a transmission connector.*

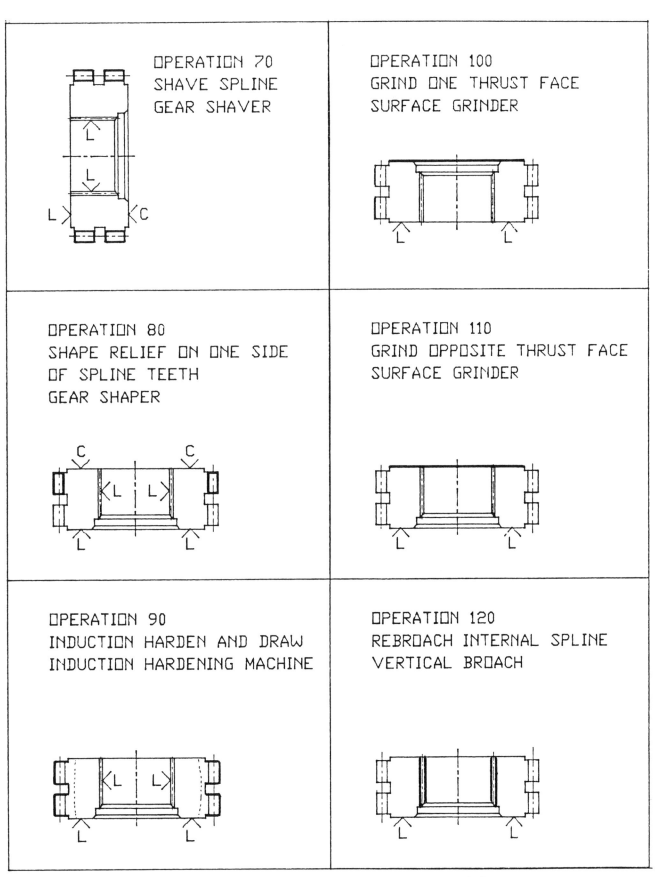

Figure 11-27. Continuation of the process planning for transmission connector.

and shaved. Shaving is necessary to ensure a good surface finish on the spline teeth. In assembly, a sliding coupler is engaged and disengaged axially with the external spline teeth. A fine surface finish is, therefore, required for smooth shifting.

Operation 80 (Figure 11-30) is a special shaping operation to make a feature that prevents jumping out of gear. One tooth flank of the upper spline is cut 0.006 to 0.010 thinner with a very light shaping cut. When coupler and connector are engaged under load, this small step between upper and lower splines will lock the mating splines together. The lock prevents jumping out of gear.

In the induction hardening operation, only the external spline teeth are hardened to below the root diameter. Dimensional variation occurs in the hardening process. This variation necessitates finish broaching of the internal spline in operation 120 (Figure 11-30).

Both thrust faces are finish ground on a surface grinder with a magnetic table. After grinding, the parts are demagnetized and washed to prevent contamination of the transmission.

Example 10: Transmission Main Shaft

The rough material for this transmission main shaft (Figure 11-28) is a cold extrusion with closely held tolerances and minimum stock allowance

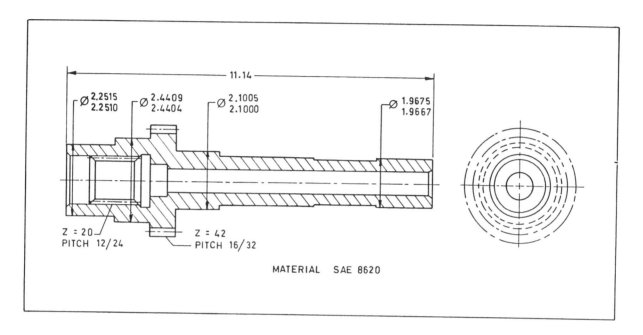

Figure 11-28. Transmission main shaft.

for turning. The higher cost of the extrusion is offset by the elimination of a gundrilling operation for the central hole through the shaft. The hole is extruded within part print tolerances and does not require further machining. The internal diameters are preformed in the shaft leaving only light metal removal in operation 30 (Figure 11-29).

The outside diameter of the external spline is ground in operation 40 (Figure 11-29). It is used as location diameter in operation 50 (Figure 11-29) to shape the internal spline.

The external spline is finish rolled with the Roto-Flo method in operation 70 (Figure 11-30). Rolling of splines on a hollow part is possible, provided the wall thickness is at least 0.375 in. (9.9 mm). The spline must also not be located near the end of the shaft, as rolling will result in a tapered condition.

To ensure good quality in the grinding operations, the part is straightened after heat treatment and the centers are ground. In a thin-walled tubing like this, the centers are likely to have ovality after heat treatment. The ovality causes roundness problems and taper of ground diameters. Ovality must therefore be corrected by a center grinding operation.

The external bearing diameters are ground in two angle head grinding operations and one operation on a cylindrical grinder. Two superfinishing operations are planned to hold the microfinish on needle bearing diameters below 16 microinches (0.4 μm).

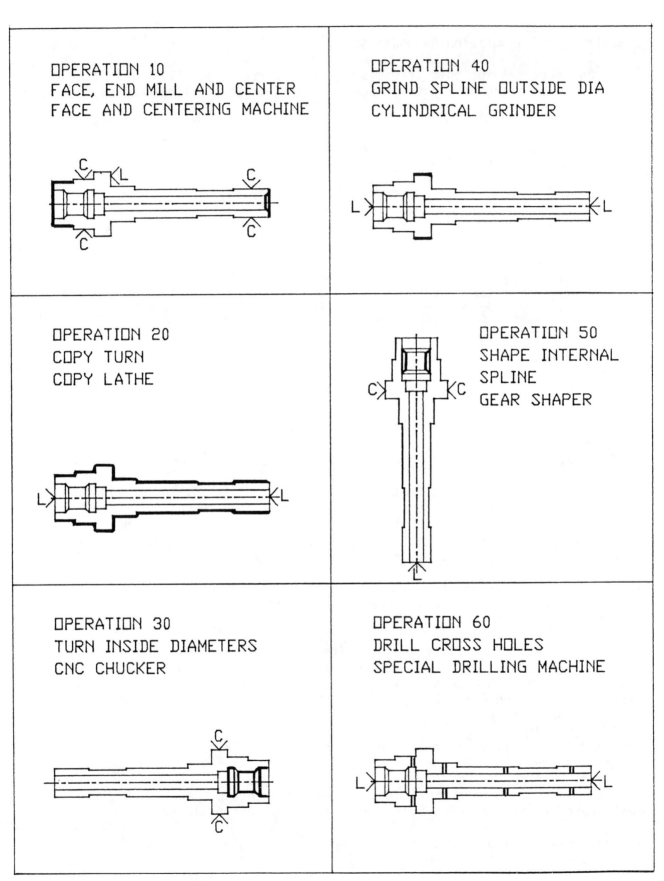

Figure 11-29. Process planning for a transmission main shaft.

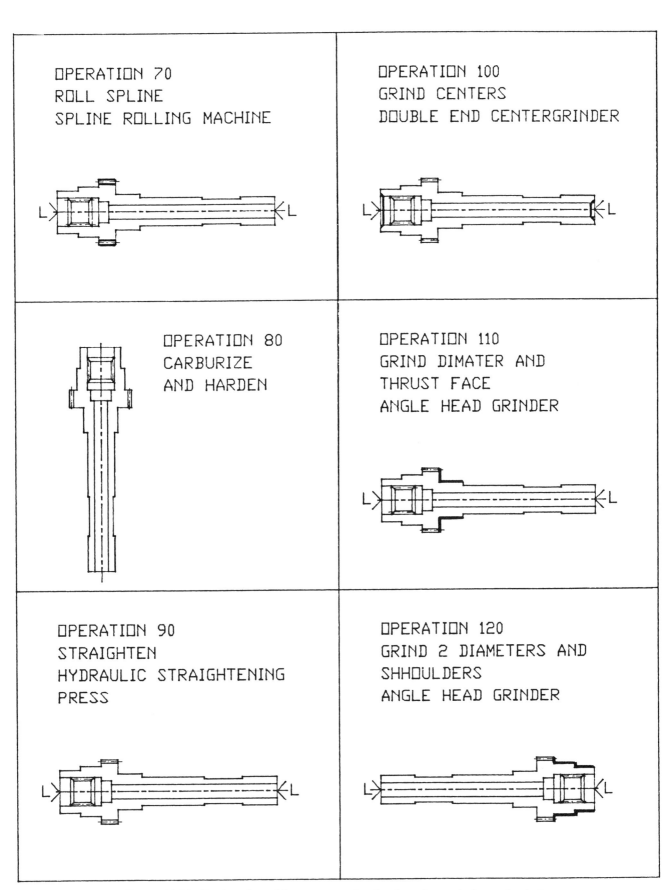

Figure 11-30. *Continuation of the process planning for a transmission main shaft.*

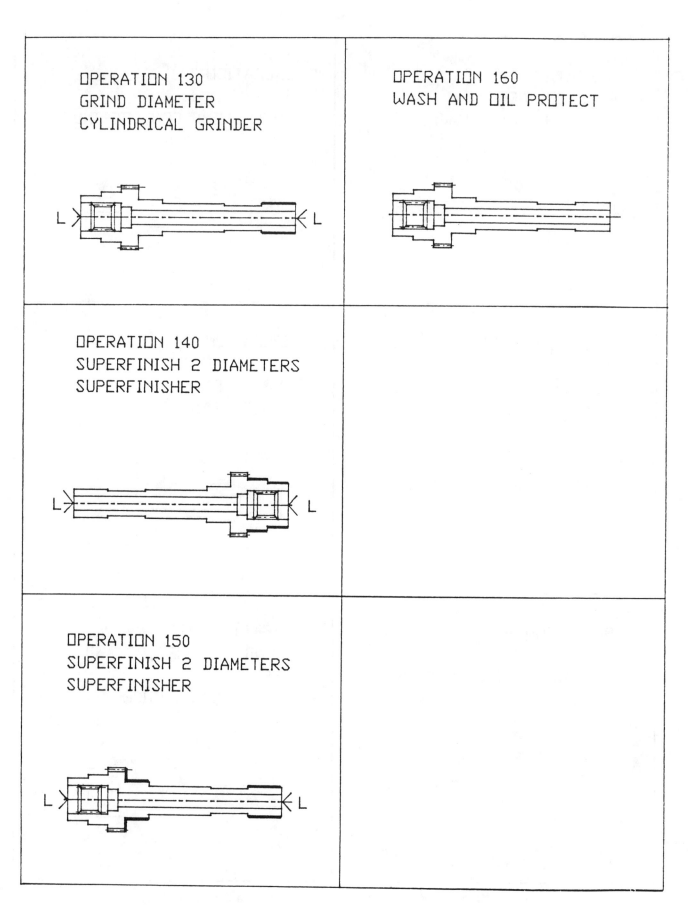

Figure 11-31. Continuation of the process planning for a transmission main shaft.

12
Gear Terminology

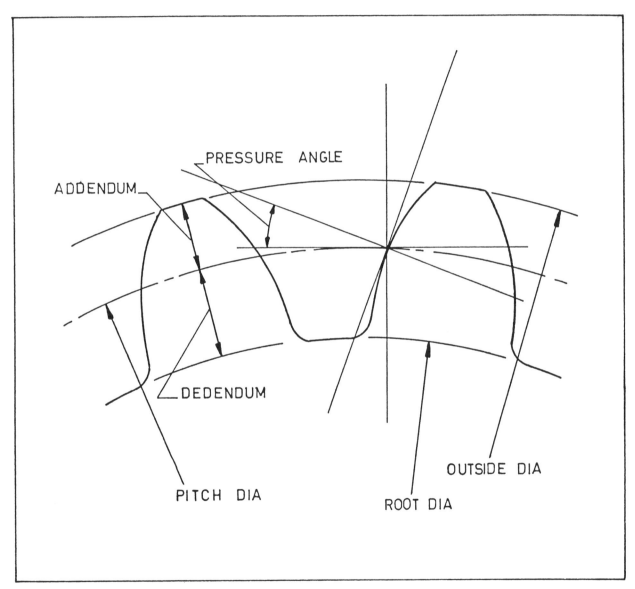

Addendum

The radial distance between the pitch circle and the outside diameter of the gear is referred to as an "addendum."

$$\text{Addendum} = \frac{1}{\text{Diametral pitch}} \text{ (inch)}$$

Circular Pitch

"Circular pitch" is the distance along the pitch circle between corresponding profiles of adjacent teeth.

$$\text{Circular pitch} = \frac{3.14159 \times \text{Pitch diameter}}{\text{Number of teeth}}$$

Dedendum

The radial distance between the pitch circle and root circle is called the "dedendum."

Diametral Pitch

This type of pitch refers to the ratio of the number of teeth to the number of inches in the pitch diameter. Metric gears use the module system to indicate tooth size. The relationship between module and diametral pitch is expressed as:

$$\text{Module} = \frac{25.4}{\text{Diametral pitch}} \text{ (mm)}$$

Helix Angle

The "helix angle" is located between a tangent to the helix and an element of the cylinder.

Lead Angle

The angle between a tangent to the pitch helix and a plane of rotation is known as a "lead angle." The lead angle is the complement of the helix angle and is mostly used to designate worms.

Outside Diameter

The "outside diameter" is the diameter of a circle measured over the top of the gear teeth. In splines the outside diameter is called "major diameter." In standard gears the outside diameter is equal to the pitch diameter plus 2x (addendum).

Pitch Diameter

The "pitch diameter" is a calculated theoretical diameter of the circumference of the number of teeth multiplied by the pitch. This is expressed by the formula:

$$\text{Pitch diameter} = \frac{\text{Number of teeth}}{\text{Diametral pitch}}$$

Pressure Angle

A "pressure angle" is the angle between a tangent to the tooth profile and a line perpendicular to the pitch surface. Usually, in gears, the pressure angle is equal to the pressure angle of the cutters used to generate the teeth.

Root Diameter

The "root diameter" is the diameter of the circle measured over the root of the gear teeth. In splines, the root diameter is called "minor diameter."

Root Relief

A modification of the involute profile where a small amount of material is removed near the root of the gear tooth is known as "root relief."

Tip Relief

A modification of the involute profile where a small amount of material is removed near the top of the gear tooth is called "tip relief."

Whole Depth

"Whole depth" is the radial distance between root diameter and outside diameter.

Part III

Bibliography

De Belder, F., Kempeneer, L. <u>Het Slijpen in de Metaalverwerkende Nijverheid,</u> Ford Tractor Belgium, Ltd, 1986.

Drozda, T. J., Wick, C. "Gear and Spline Production," <u>Tool and Manufacturing Engineers Handbook</u> (Fourth Edition, Vol. 1), Dearborn, MI: Society of Manufacturing Engineers, 1983.

Ex-Cell-O Corporation, "Michigan Shear Speed Gear Shapers" Bulletin MSS-67, 1967.

Ex-Cell-O Corporation, "Process Systems Operations," Detroit, 1982.

"Fellows Gear Shaper Cutters," Springfield, VT: Fellows Corporation, 1975.

Hoffman, K. <u>Hoffman Raumpraxis,</u> Pforzheim, 1976.

Keck, K. F. <u>Getriebe Atlas I/1,</u> Ludwigsburg, 1966.

Kelly, P. W. "Advantages of Titanium Nitride Coated Tools" <u>The Journal of Gear Manufacturing,</u> May/June 1984.

Lange, J. M. "Gear Shaping Machines CNC Development " <u>The Journal of Gear Manufacturing,</u> November/December 1985.

Moens, K., Vandenhende, U. <u>Het Vervaardigen van Tandwielen,</u> Ford Tractor Belgium, Ltd., October 1985.

National Broach & Machine Division, <u>Modern Methods of Gear Manufacture,</u> Detroit, 1972.

National Broach & Machine Co. "Broaching Practice," Detroit, 1953.

National Broach & Machine Co. "Design and Application of High Speed Broaching Tools," Detroit, 1978.

Pfauter Letter 45, "Skiving of Internal Gears," Ludwigsburg, 1974.

Pfauter Letter 47, "Carbide Finish Hobbing" Ludwigsburg, August 1981.

Psenka, J. A. "Producing Gears by the Broaching Process" AGMA Conference, Chicago, 1961.

Rice, C. "The Cost of Gear Tooth Accuracy," SME Conference: Gear Processing & Manufacturing, December 1977.

Sulzer, G. "CNC Gear Shaping" <u>The Journal of Gear Manufacturing,</u> March/April 1986.

The Fellow Gear Shaper Company, "Fellows Equipment for Cutting, Finishing, and Testing Gears." Springfield, VT: 1955.

Index

A

Addendum, 173
Angle head grinding, 133-138
Annealing, 119
Automated part handling, 108

B

Barstock, 107
Blanking, 105-117
Bore grinding, 134
Brinell Hardness, 51
Broach cycles, 87

C

Calculation of hob travel, 29-30
Carbide finish hobbing, 88
Carbonitriding, 120
Carburizing, 120
Case hardening, 118
Centerless grinding, 132
Chamfering, 101-102
Circular pitch, 174
Climb hobbing, 9, 44, 45
Cluster gears, 142-147
CNC, See: Computer numerical control
Combination shaping, 57
Computer numerical control, 6, 41, 108, 131, 133
Conventional hobbing, 9-10, 45
Conventional shaving, 63, 67
Countershaft design, 123
Countershafts, 123, 148-150
Crossed axis angles, 21
Crown hobbing, 44, 46
Crowning, 44, 46, 73
Cutter sizes, 65
Cutting speeds, 35, 50, 52
Cycle calculations, 37
Cycle time elements, 49-61
Cycle times, 28, 37, 39, 42, 64
Cylindrical gears, 78
Cylindrical grinding, 130-132

D

Deburring, 101-102
Dedendum, 174
Design, 123
Diagonal shaving, 63, 67, 72
Diametral pitch, 59, 174
Dimensional tolerances, 127
Distortions, 121-125
Double cut hobbing cycles, 28
Down shaping, 59

E

External gear broaching, 87-88
External gears, 60, 87-88
External recessed tooth forms, 60
External spline broaching, 96
Extrusions, 109-110

F

Face grinding, 133, 134
Feeds, 49, 51
Flame hardening, 121
Floor layout, 108
Forgings, 109
Form grinding, 82-83
Form tolerances, 129

G

Gear face width, 29
Gear honing, 85
Gear milling, 85
Gear rolling, 82
Gear shaping machines, 18-19
Gear sounding, 103-104
Grinding
   angle head, 133-138
   bore, 134
   centerless, 132
   computer numerical control, 131
   dimensional tolerances, 127

disc, 128
  face, 133, 134
  finish, 83
  form 82-83, 129
  form tolerances, 129
  generating, 85
  groove, 132
  horizontal disc, 128
  internal, 134
  multiple diameters, 133
  multiwheel, 131
  peripheral, 137
  plunge, 130
  recessed face, 136
  surface, 127-128, 137
  traverse, 131
  vertical disc, 137
Groove grinding, 132

H

Heat treating, 117-125
Helix angles, 174
Hobbing
  applications, 5, 43-48
  carbide finish hobbing, 88-89
  centering, 7
  climb, 9, 44, 45
  crown, 44, 46
  cycle calculations, 37-43
  cycle time formulas, 28
  double cut, 28
  gear face width, 29
  machines, 6
  number of starts, 36
  overrun, 32
  process defined, 27-37
  shifting, 7-8
  single cut hobbing, 28
  spacer width, 30
  spline, 95
  taper, 44, 46
  travel, 29-37
Hob centering, 7
Hob construction, 12
Hob revolutions, 34
Hob shifting, 7-8
Hob starts, 36
Hob travel, 29-20
Horizontal disc grinding, 138

I

Idler gears, 139-141
Induction hardening, 118, 121
Industrial robots, 108

Infeed times, 49
Internal gear broaching, 86-87
Internal gears, 60, 85, 86-87
Internal gear skiving, 85
Internal planetary ring gears, 151-152, 162-164
Internal spline broaching, 95-96
Involute modifications, 74
Involute spline grinding, 98
Involute splines, 91, 98

L

Labor, 108
Lead angles, 174
Length of stroke, 52
Low carbon steels, 118

M

Machining, 29, 117, 122
Machining times, 29,
Material handling, 108
Medium carbon steels, 118
Milling, 93
Multiwheel grinding, 131

N

Nitriding, 120
Normalizing, 119

O

Outside diameter, 174
Overrun, 32

P

Part configurations, 77-78, 81, 89, 105
Peripheral grinding, 137
Pitch diameter, 174
Plastic deformation, 122
Plunge grinding, 130
Plunge shaving, 63
Potbroaching, 93
Pressure angles, 175
Production volumes, 80-81, 92, 106
Profile, 15
Progressive blind broaching, 96

Q

Quality
  class, 12, 14
  levels, 79

specifications, 79-80

R

Recessed face, 136
Reciprocating cutter speeds, 50
Robots, 108
Rolling, 93
Root diameters, 175
Root relief, 175
Rotary feed times, 49
Rough material, 107

S

Shaper cutters, 19-21
Shaping
  applications, 56-57
  basis of, 49
  combination shaping, 57
  cutting speeds, 50-52
  cycle time elements, 49-60
  down shaping, 58
  gear machines, 18-19
  length of stroke, 52-53
  machine, 18-19
  shaper cutters, 19-20
  spline manufacturing, 91, 95
  taper shaping, 57
  three cuts cycle, 54-56
  two cuts cycle, 53-54
Shaving
  applications, 73
  crowning, 73
  cutting feed, 65-69
  cutters, 25
  cutting speed, 64
  cycle time calculations, 71-73
  diagonal, 63, 67, 72
  involute modifications, 74
  machines, 24-25
  methods of, 63
  number of strokes, 69
  purpose, 21
  spline, 95
  stroke length, 66-67
Shear cutting, 93, 96
Single cut cycles, 28, 53
Single cut hobbing cycles, 28
Spacer widths, 30

Spline grinding, 97
Spline hobbing, 97
Spline manufacturing, 89-99
Spline milling, 95
Spline rolling, 92
Spline shaping, 97
Spline shaving, 97
Straight sided spline grinding, 98
Straight sided splines, 91
Stroke length, 66, 67
Surface enrichment, 118
Surface finish, 127
Surface grinding, 137-138
Surface roughness, 128
Synchronized transmissions, 158-161

T

Taper hobbing, 44, 46
Tapers, 44, 46, 57, 73
Taper shaping, 57
Three cut cycles, 54
Tip relief, 175
Tolerances, 129
Tool forms, 89-90
Tooling, 108
Tool materials, 16
Toothform, 12
Tooth pointing, 102-103
Tooth rounding, 102-103
Transmission connectors, 165-168
Transmission countershafts, 114, 153-157
Transmission main shafts, 169-172
Traverse grinding, 131
Two cut cycles, 53

U

Underpass shaving, 63
Universal tooling, 108

V

Variable feeds, 47
Vertical disc grinding, 137

W

Whole depth, 175